QH60   Harris
H36   Natural history collecting

DATE DUE			
OCT 29 '80			

A
GROSSET
ALL-COLOR GUIDE

NATURAL HISTORY COLLECTING

BY REG HARRIS
Illustrated by Peter Thornley

GROSSET & DUNLAP
A NATIONAL GENERAL COMPANY
Publishers · New York

CONTENTS

4 **Introduction**

8 **Kinds of Collections**

12 **Collection Apparatus**

24 **Collecting Plants**

44 **Collecting Animals**

44 Invertebrates

77 Vertebrates

78 Fishes

87 Amphibians

94 Reptiles

103 Birds

120 Mammals

138 **Collection Areas**

143 **Collecting Fossils, Rocks and Minerals**

149 **Setting Up Your Collection**

156 **Books to Read**

157 **Index**

The Hanging Gardens of Babylon

INTRODUCTION

Man is an inquisitive creature and it is not surprising that he has always been interested in collecting objects of various kinds, particularly those relating to natural history.

From the earliest times man has picked up and kept the odd rock, skull or bone that attracted his attention and it is probable that from these pieces arose the idea of collecting objects to compare them with each other. The Roman Circus was a somewhat short-lived attempt at animal collection, for despite the mortality of animals collected for circus entertainment, many rare and unusual animals were discovered and exhibited.

One of the seven wonders of the ancient world was the Hanging Gardens of Babylon. These so-called hanging gardens were built up to the height of the city wall on arches, of which very few have survived to the present day. On these arches terraces were built, on which soil was laid so that trees and shrubs could grow.

The earliest reference to a botanical garden in Mesopotamia is in inscriptions attributed to Tiglath-Pileser I (1100 B.C.). After him, it was the custom of rulers to carry on the building and improvement of the gardens. A description from writing tablets of Ashurnasirpal II (875 B.C.) says 'From the countries in which I traveled, and from the mountains that I passed, I planted trees and seeds of cedar, cypress, box, pine, date palm, oak, tamarisk, laurel, poplar, willow, pomegranate, medlar, pear, quince, aloes and sycamore.' These botanical gardens were used as government experimental stations.

There are many botanical gardens in North America which are open to the general public and are worth a visit.

An impression of a Roman Circus, where many animals were exhibited and killed for the enjoyment of the people.

The early alchemists had collections of plants and animals as part of their paraphernalia. This led, quite naturally, to the first collections to be used for teaching purposes. By this time men were interested in collecting specimens for curiosity rather than for any special reason, and this kind of collecting habit is still prevalent. It is probably true to say that the mania for collecting of the Victorian Age will never again be repeated, but nevertheless there is a place for the collection of natural history objects and for their care and conservation. It is an inexpensive pastime to look for the odd and unusual in nature and it is a most rewarding experience.

Frederich Ruysch of Amsterdam (1638–1731) was one of the most interesting of early collectors. His displays were artistic in a most peculiar fashion and would often include fossils, rocks, plants of many varieties and embryonic stages of animals, including man. Ruysch's preparations were ar-

An early alchemist at work.

ranged decoratively but did not show any specific biological principle. A pile of stones from the kidney (Renal calculi) were often the base onto which trees of dried blood vessels were built. A dried skeleton clasping a mayfly and a series of mollusk shells were also included, and jars of preserved fish were also used. By 1710, he had made 1,300 fluid-preserved preparations including a considerable botanical collection. Peter the Great of Russia bought this huge collection that Ruysch had amassed, and when the collection arrived in St. Petersburg (now Leningrad) it was found that the sailors had drunk all the brandy in which many of the specimens had been preserved. Although Ruysch was then over 80 years of age, he set about creating a new collection.

One of the most interesting of the eighteenth century collectors was the Surgeon General to the British Army, John Hunter. Qualifying as a doctor, he spent a good deal of his time on experiments in blood circulation and other physiological work. He was an extremely nervous man, and it is recorded that he had to take a dose of laudanum before giving a lecture. He founded a unique collection of natural history specimens known as the Hunterian Collection which, in its heyday, contained over 13,000 specimens. Many of these were destroyed during World War II. It was Hunter's conception to turn a museum from a series of curiosities into an ordered, instructive and useful collection. This collection may still be seen at the Royal College of Surgeons, in London.

Frederick Ruysch (1638–1731) formed artistic displays of rocks, pebbles, dried blood vessels, fossils and embryonic stages of animals, including man.

National parks are areas set apart for the preservation and observation of wildlife and the natural environment.

KINDS OF COLLECTIONS

There are many kinds of nature collections. The largest and most organized are attempts to present animals and plants in a natural or semi-natural state. Examples of these kinds of collections can be found in the national parks, national forests and national wildlife refuges. Perhaps best known of these in the United States is Yellowstone National Park. Many states also have forests and parks where they attempt to conserve natural resources and, at the same time, enable people to get a closer look. Zoological and botanical gardens attempt more and more to present their collections in as natural a setting as possible. Preserved natural history material is displayed in public museums throughout North America.

Collecting Plant Life

Flowers may be collected and preserved in season, many of them retaining their color and detail. Collections of plants of similar habitat may be made. Climbing plants, insectivo-

Beneath the huge aluminum and glass domes at Milwaukee's *Climatron*, visitors can view vegetation in man-made climatic zones.

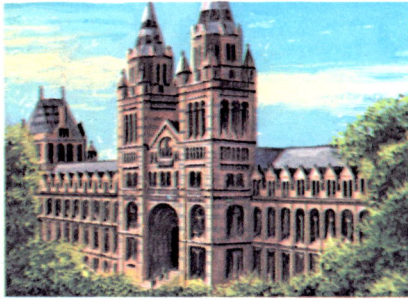

Museums are not only 'storehouses' for things of the past, but also active places of scientific research.

rous types and cacti are particularly interesting specimens for collections. Grasses of various sorts may be pressed and mounted.

Collecting Animal Life

The many different species of insects such as beetles, butterflies and moths can be caught, killed with special fluids, dried, mounted on pins and arranged in glass-topped boxes to form very fine collections. These creatures, however, can also be observed and records made of their eating habits—to see whether they feed on animal or vegetable material—their way of life and so on. Insect larval forms, especially caterpillars, can be dried and mounted together with the adult stages to show how insects develop.

Skulls of various animals, teeth and bones are also interesting objects to form collections for comparison purposes. The contents of the food pellets of owls, herons and other predatory birds can form the basis of collections of bones, skulls and general skeletal parts.

The Great Flight Cage at the National Zoological Park in Washington, D.C. enables visitors to view birds at a close range.

9

On the seashore, plants and animals of various kinds and with their own particular mode of life, according to the nature of the shore, will be found in great profusion.

In many cases, photographic records will suffice and, in fact, present fewer storage problems than actual specimens. Besides being a more conservation-minded type of collection, such a collection is more adaptable for wider use — exhibits and displays, lectures, illustrations.

Collecting Rocks, Minerals and Fossils

Collections of objects of geological interest can easily be assembled. Rocks are of three kinds: igneous, sedimentary and metamorphic.

Igneous rocks are of volcanic origin and are both coarse and fine-grained in structure. Examples include lava, pumice and granite. Sedimentary rocks were laid down in water millions of years ago. They include sandstones, clays and limestone, which produces bubbles of carbon dioxide gas when vinegar is poured on it.

Fossils

Beetle

Wood Anemone

Shell

Metamorphic rocks were formed by the action of heat and pressure and have a variety of crystalline forms.

Rocks and minerals are found in many different areas, and the varied nature of these objects gives an opportunity to spend many interesting leisure hours. The seashore and lakeside are also well worth examining for the wide variety of plant and animal life, and pebbles and rocks.

Minerals are colorful stones to collect, and can be found in ore dumps and in cliffs and on the seashore. Old mines may contain minerals, but they are dangerous places to enter. The appearance of pieces of mineral can be enhanced by polishing them in a rotating tumbler containing other stones and water.

Fossils are found in quarries and in cliff faces and along the seashore. They should be collected in groups such as ammonites, fish teeth and bones.

The identification, classification and exhibition of fossils can be done back in your home, and it is an especially pleasant pastime on winter evenings.

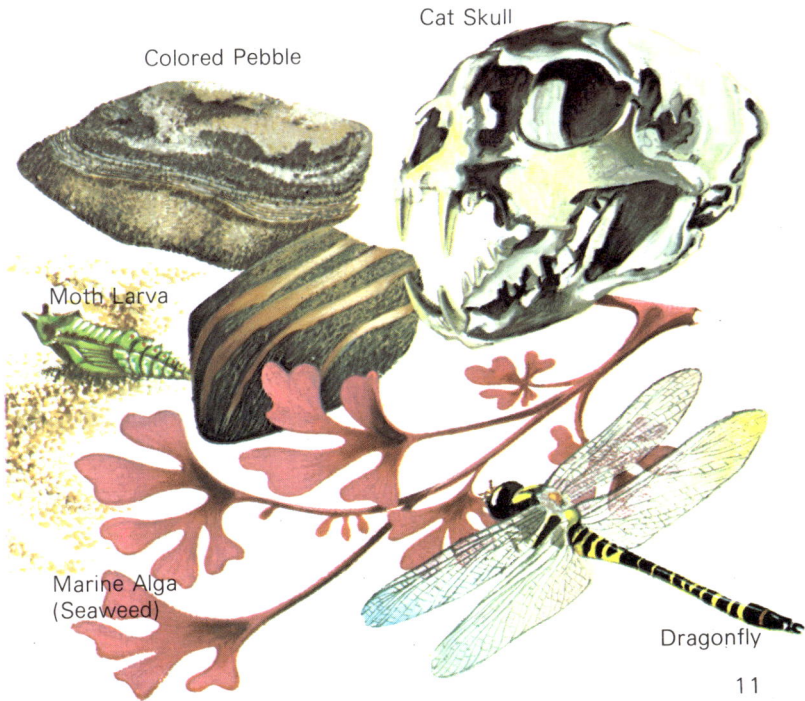

Cat Skull

Colored Pebble

Moth Larva

Marine Alga
(Seaweed)

Dragonfly

11

COLLECTION APPARATUS

Plants

A small plant press for collecting specimens in the field may be constructed from two sheets of plywood or from perforated zinc sheets and galvanized iron wire. The perforations allow air to reach the plants, which are placed between folded sheets of absorbent paper. Sheets of newspaper are adequate for this purpose. Blotting-paper is not suitable because it absorbs the moisture too quickly, and repeated damping and drying would render the blotting-paper useless. Place each plant singly on a sheet if possible, although small ones may be laid side by side, and fold the paper over them. This process is repeated for the next specimen, and

A portable plant press

A vasculum, used to carry collected plant specimens

each sheet is laid with the folds alternating so that a regular and level height is maintained. To exert pressure on the plants, two or more canvas straps, string or cord are tied around the frame, and for extra pressure a heavy stone or a series of books may be placed on top.

The paper should be changed every two or three days until the plants are dry. If the procedure is carried out with care, very good results can be expected. Succulent plants may have to be plunged into hot or boiling water for a few minutes before pressing, in order to kill the tissues and accelerate the drying. Without this treatment, they may develop molds during the drying period.

A small book will serve as an efficient plant press. It is excellent for pressing small leaves or flowers collected during a walk in the country. Place a few sheets of thin blotting-paper between the pages to encourage quick drying.

A botanical vasculum is the ideal container for carrying collected plants. They are obtained in various sizes and are best made from aluminum, although old ones were made from tin plate. These vascula will not rust and will withstand the hardest treatment. Aluminum vascula throw back heat and remain cool, whereas the old familiar black tin vasculum absorbed much heat.

Animals

Several small glass or plastic tubes, some wide-mouthed jars, a supply of plastic bags of various sizes and some plastic sheeting in which to keep specimens are required. These bags will prove useful for keeping plants in. Small tubes or bottles may be used to gather samples from roof gutters, ditches and puddles. It is surprising the amount of material able to be collected in this way. A small hand-lens will increase the amount of visible material enormously.

Nets

Many nets are available for collecting purposes. One can easily be made by attaching part of an old nylon stocking to a wire frame fixed to a bamboo cane.

Beginners are advised to start with the simplest and minimum of equipment, and to increase their apparatus as the need arises from experience gained.

Experience in the use of nets will be necessary when attempting to collect specified material. For example, when collecting fast-moving organisms such as water beetles, it is advisable to pass a coarse-meshed net quickly through a batch of weeds and then to deposit the net contents quickly into a plastic bowl or bucket. This method is easier than stalking and capturing individuals.

When a net is used, it is important to empty it as soon as possible. This will prevent the organisms from adhering to the net walls and being retained within the meshes. The

14

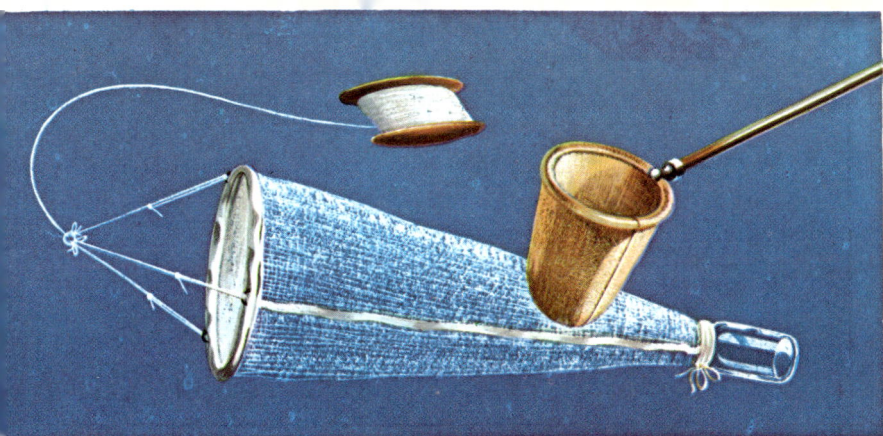

These hand nets are used to collect water creatures.

The nets with a tube at their base are used to collect plankton.

small tube usually placed at the end of most water nets aids the concentration of the collection and identification.

Water nets exist in a range of shapes, and it is usual to have several shapes with detachable frame rings to fit one pole. Small plankton nets are usually made in silk or nylon with a small container, usually a plastic tube fitted at the end of the net, for collecting the catch. Mesh sizes are available from 62 to 125 to the inch. Nets for collecting minute organisms have a mesh size of 180 to the inch, but it is unlikely that the amateur collector will need such fine sizes. The usual plankton netting has a mesh size of 62 to the inch.

Dry nets are used for butterfly and moth collections. A variety called a 'sweep' net, shaped like a frying pan, is used for beating bushes and shrubs to collect insects attached to leaves and branches. These triangular-frame or circular frame nets are made of mesh cambric or nylon with a leather protective collar over the edge of the frame to prevent wear on the net mesh.

Butterfly nets are usually made these days in terylene or nylon, although muslin is also used. Usually the net is long and pocket-shaped so that the specimen may be trapped and yet not crushed while being examined. It takes much practice to catch butterflies and moths in flight.

Spring and early summer are good times to go collecting as the areas are usually teeming with life at these times of year.

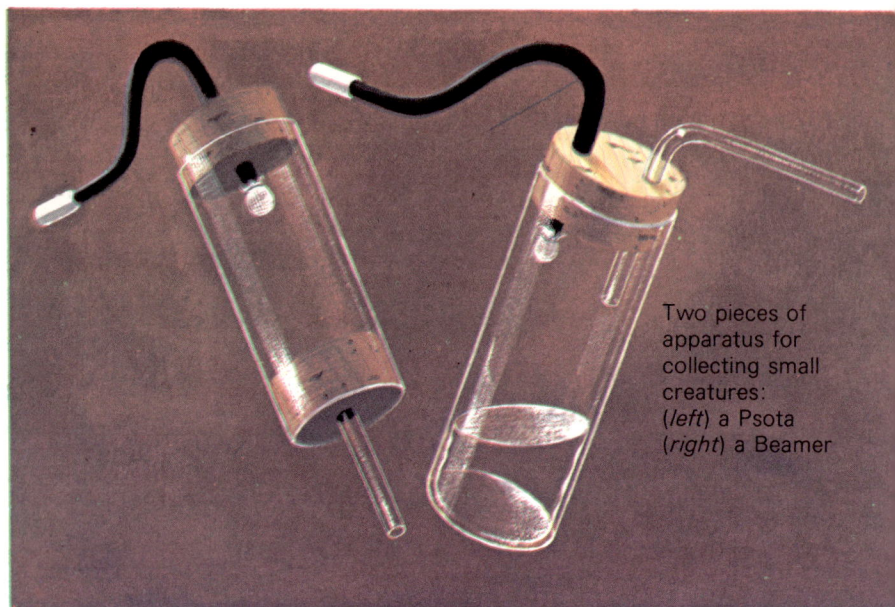

Two pieces of
apparatus for
collecting small
creatures:
(*left*) a Psota
(*right*) a Beamer

Aspirators

A pooter or aspirator is a good instrument for collecting insects and small animals. Two kinds are available: the Psota and the Beamer. The instrument is held in the hand and the specimen sucked into the tube. A piece of gauze on the end of the mouthpiece tubing prevents accidental swallowing! The Beamer-type usually has a small supply of preservative (70 percent alcohol) at the bottom of the container if immediate preservation is required.

Killing jars

A killing jar is useful for collecting insects: it is simply a wide-mouthed jar containing a suitable killing agent. Potassium cyanide is used by professionals, but it is far too dangerous for amateurs to handle and there are several other far less harmful reagents that can be used. Ethyl acetate, chloroform or ammonium hydroxide are much safer. The chemicals may be poured onto a pad of cotton wool and then placed in the jar, or a cork can be prepared for the jar and hollowed out to take the cotton wool and killing reagent.

Alternatively, a layer of plaster of Paris can be poured into the jar and when dry a few drops of the chemical added.

Chemicals to kill insects can be bought from biological supply houses. These fluids are often sold in two strengths, one for beetles and the other for butterflies and moths. The killing fluids also contain relaxing chemicals which assist easy manipulation of the insects' limbs, wings and feelers when setting them in place with pins.

Not all insects are killed by the same concentration of proprietary killing fluid or by the same length of time.

Spreading Boards

Butterflies, moths and dragonflies are mounted with their wings spread out. In order to accomplish this, they must be dried on a spreading board. A spreading board is made of two strips of soft wood, nailed parallel, with an opening between them as wide as the body of the insects to be pinned. A pinning strip of cork, balsa wood or cardboard is tacked under the groove (*see page 153*).

The insect should be pinned through the abdomen and the pin stuck into the pinning strip so that the body fits in the groove. The wings should be placed in the position desired when dry and held in place by thin strips of paper which should be pinned in place. Allow several days to dry.

Three methods of adding solutions to jars in which insects are killed: *extreme left*, impregnated cotton wool in base; *center*, killing fluid in cork; *right*, impregnated plaster in base

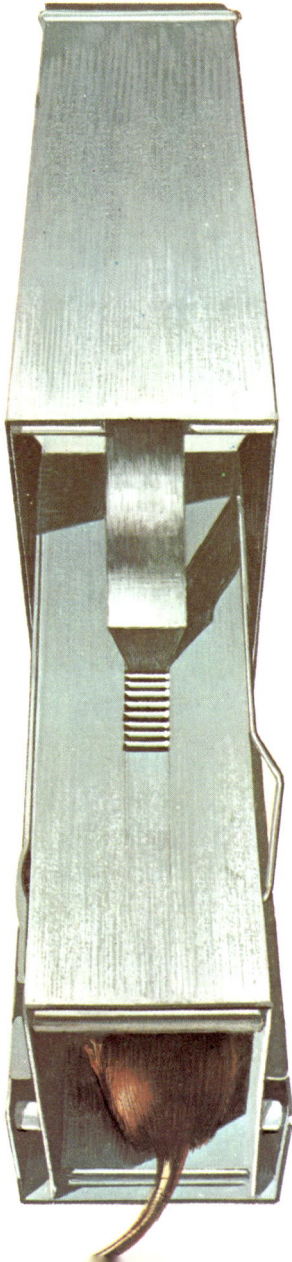

A trap used for collecting many small mammals. Use gloves when handling the animals

Small Mammal Traps

Small traps are obtainable for the capture of small mammals without harming them. They are made of aluminum to prevent corrosion and deterioration, and are usually set in obvious animal runs in fields, woodlands and along pond edges. They are left for a period of time varying from a few hours to overnight. It is good practice to visit the traps at regular intervals to see if the trap door has been sprung. In most cases the small mammal settles down quite well once captured, especially if the trap has been baited with some food. However, if a shrew has been caught, it is important not to leave the little creature too long; it has an extremely nervous disposition and might otherwise die of shock after making frantic efforts to escape. After the captured animal has been placed in a container for observation and possibly photographed, it is released, none the worse for its experience.

As previously stated, the traps can be baited, but this is not essential as most small mammals are insatiably inquisitive, and will enter any small aperture that interests them. A good plan is to rub the trap over with cooking fat, bacon grease, or a piece of ripe cheese. The smell will block out the human scent, and is attractive to a number of small mammals. Not only these animals are caught—sometimes amphibians (frogs or toads) will find their way into the trap, as will the occasional small snake. So be prepared for anything when the trap is sprung. Be sure to check state and local laws and ordinances before engaging in any trapping activity.

Extraction Funnels

A Berlese extraction funnel is used for collecting animals from debris in surroundings such as woodlands. A tin is placed over a funnel which is supported over a small container of alcohol. A light source is placed on or near the funnel top and a small piece of perforated zinc is placed at the bottom of the funnel. The heat from the light will cause the animals to move and they will fall into the preservative.

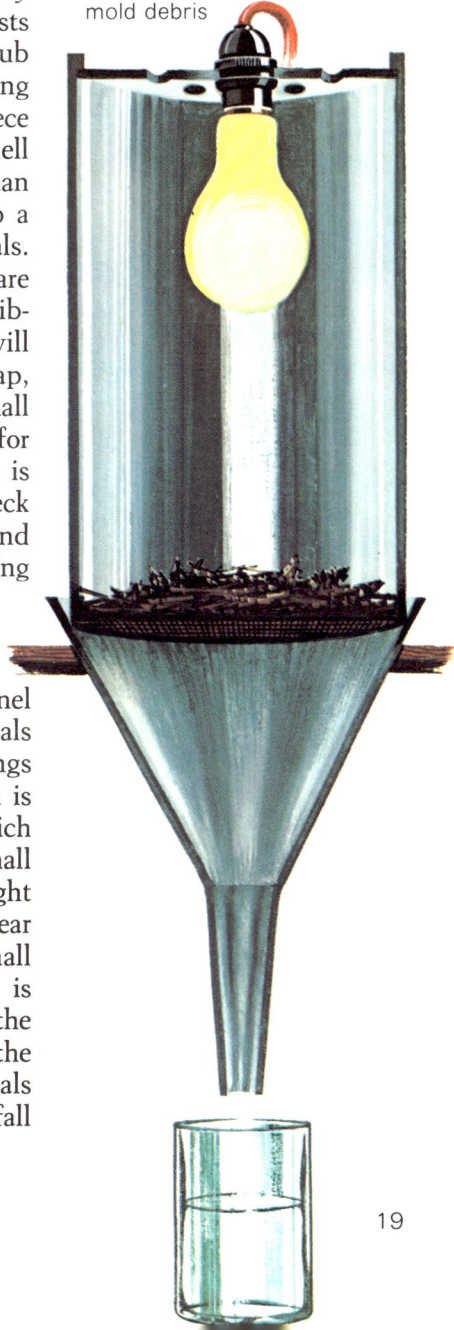

A Berlese Extractor Apparatus for trapping insects in leaf and mold debris

19

Other Useful Apparatus

Garden tools such as spades, forks and hoes are very useful for collecting specimens, both in the garden and on the seashore. Collapsible spades are extremely useful, as they can be easily transported.

For more delicate work, old household spoons and forks can be the most useful pieces of equipment. Fork prongs can be bent or removed, forming tools which enable corners and crevices to be poked into. Trowels also make handy tools for forming holes in sand and mud, from which small invertebrate animals such as sand hoppers and worms can be caught.

Rubber boots or hiking shoes are extremely useful, especially when collecting in marshy and muddy areas on the seashore. Although sneakers may be worn, and bare feet are often preferred on a sandy beach, it is a mistake to undertake prolonged collection on the seashore without wearing rubber boots. Quite apart from the cold (even in summer the sea can feel very cold after an hour or two), there is the risk of slipping on rocks and treading on debris.

For rock and mineral collection as well as for barnacle and lichen collection, a small geological hammer is required. These hammers are available in many sizes. The beginner

Gardening tools are excellent for digging up animals, especially on the seashore.

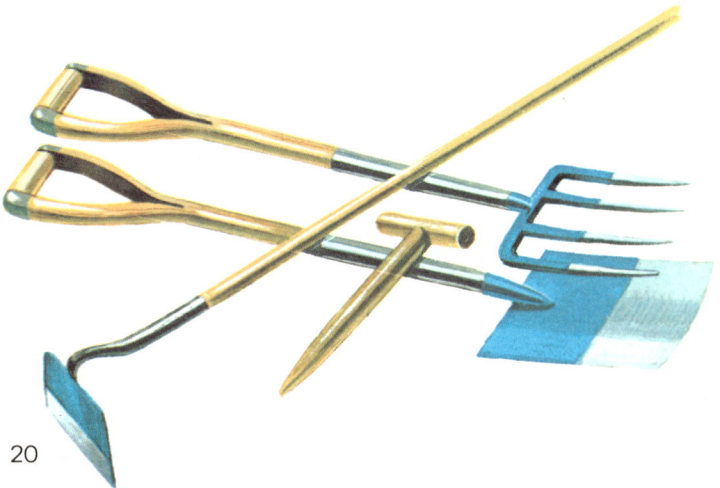

can start with a small coal hammer, with the purchase of the more expensive geological hammers at a later date.

A cold chisel is needed to break away rocks. A supply of newspapers is useful to wrap the fossils or minerals collected and to separate and to protect the more delicate specimens.

Plastic bags can be extremely useful, especially to enable wet, muddy and dirty specimens to be transported home without other articles in the knapsack becoming contaminated. These bags, however, should not be left scattered in fields, especially where animals are grazing. A plastic bag when swallowed by an animal will block air passages and tubes to the stomach, causing death. This state of affairs does not bring about a happy relationship between genuine natural history collectors and farmers.

For keeping the collection and for carrying it about, a basket divided into compartments is convenient. Jam jars or coffee jars with screw lids may be used to keep the specimens. Remember that sea-water will erode most metals in time. A supply of tie-on and sticky labels will be very helpful in sorting out the catch. For specimens collected in jars of fluid, it is advisable to place a paper label with the information written in pencil inside the container and a sticky label on the outside. This will ensure that the specimens can be identified even if the outside label is removed.

After their capture, these animals should be placed in separate jars. A hand lens aids in identification.

Preparation for Collecting Trips

Always prepare well for your collecting trips. Thorough preparation and careful thought of the equipment needed will save many wasted journeys. Few things are as annoying as to arrive at the scene of collection and then to search frantically for a bottle or jar of killing fluid which was accidentally left at home. A list of equipment required will prevent all of this trouble.

What shall I use to carry all this equipment? is the next question often asked. Rucksacks and knapsacks are the best holders. Bags which have to be put down every time one wishes to pick up an insect, pebble or flower are both time wasting and annoying. There is a wide range of knapsacks available.

When only a few items of equipment have to be transported, a frameless rucksack is better than a framed one, as it enables the rucksack to be slung over a shoulder.

A hand lens attached to a length of string and hung around the neck is helpful for preliminary identification. These hand lenses are usually groups of three lenses in a metal or plastic case with a small eyehole through which string or cord can be threaded.

Remember to take a small pocket notebook with you on your collecting trip. All information is useful, even the time

Waterproofed boots, newspapers, chisels and geologists' picks are useful in collecting fossils.

22

Various kinds of packs provide an easy way to transport paraphernalia for nature collecting.

of day that the collection was taken and the state of the weather. If on the seashore, note the state of the tide. If collecting fossils and minerals, record as much information as possible about the area, quarry, cutting or working that you can obtain. Do not forget to take a small first-aid kit with you, as a cut hand can easily be infected when handling animals and plants.

A good, accurate map is essential. Detailed topographical maps can be obtained for local areas from the U.S. Department of the Interior in Washington, D.C. Often, and especially in wet weather, the map can be opened at the section required and placed in a plastic bag. This method saves crouching near a wall or groping beneath a rain-sodden raincoat, when attempting to locate your position on a map during a storm.

Safety Precautions
If, by chance, your collecting trip takes you into wild, uninhabited country, remember to tell a reliable person or someone in authority where you are going and what time you expect to return. Sudden storms or fog can soon bring danger in remote areas.

Luckily, most areas for collection are perfectly safe, but never take chances at cliff edges. Remember that it is best not to go alone, but to have at least one companion.

COLLECTING PLANTS

There are four major groups of plants. One of these groups is the Thallophyta, which include the bacteria; the algae, which are represented by the seaweeds and by many small fresh-water plants; the fungi, which include toadstools and molds; and the lichens, which are actually algae and fungi living together for mutual existence. Others are the Bryophyta, which include the mosses and liverworts; the Pteridophyta, including the horsetails, ferns and club mosses; and the Spermatophyta, which include the seed plants or flowering plants.

Plants are formed in collections known as herbaria. One of the largest in the United States is the New York Botanical Garden in the Bronx. These collections of dried plants have a

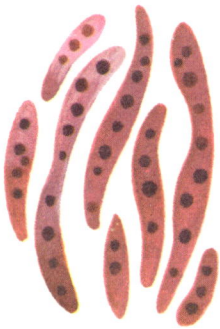

Bacteria   Fungi   Fungi

Moss   Liverwort   Ferns

very important function in economic research medicine and pharmacy. Many important and useful facts have evolved from the study of a herbarium sheet. Plant enthusiasts can easily make a small, dried collection using cardboard folders to contain the sheets on which the plants are attached. Some plants are too bulky for this sort of treatment. These are dried and placed in small boxes.

Some plants may be carefully dried and retained in their natural shape and position. Many of the tall swamp grasses, common on the edges of streams and rivers and slow flowing canals, can be collected in the summer and allowed to dry slowly in a tall jar without water. The author has had sheaves of tall grasses for several years, and they form an extremely attractive sight.

Mold

Algae

Lichen

Pink

Dandelion

Violet

Pondweed

Algae

These plants vary in size from the small microscopic cells found in ponds and streams to the giant seaweeds of the open sea. The collection of freshwater algae is comparatively simple. When pond or stream dipping, a good deal of algae will usually be recovered in the net. This is spread on a piece of glass or plastic, or on a microscope slide for examination under a hand lens or a microscope.

Marine algae are collected from the rocks and seashore, or from the sea bottom with dredges from a small boat. The larger varieties can be dried out by hanging the plants on a clothes hanger in a warm place. The smaller, more fragile and often highly colored algae may be spread out on glass or plastic, as mentioned for the pond examples, and allowed to dry naturally in a warm place.

Probably the most important algae that a young collector might obtain are the seaweeds. Scientists usually divide

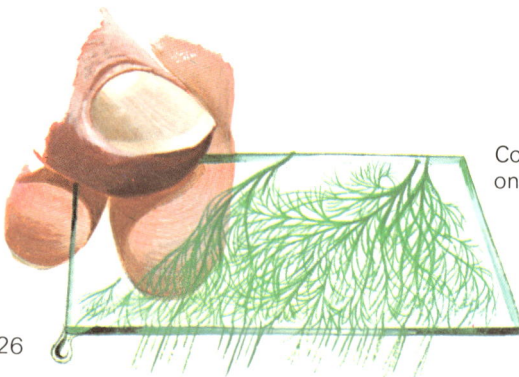

Collecting algae on a slide

Seaweeds

them into three main groups, according to their color—green, red or brown.

Many of the brown weeds may look green, and many of the red varieties may look brown at certain times, especially if they are breaking down or decomposing. The true color occurs only in fresh and actively living algae. All these plants will bleach white when dead and dried in the sun. Freshly collected algae can be dried with good results of color preservation if the color is well shown at the time of collecting and if the procedure described on the previous page is carried out carefully.

There are thousands of seaweeds off the North American coasts. It is practically impossible to tell which weed is which when there is only a few inches of plant available. There are a number of very useful books which help in the identification of these interesting plants. Museums will also often identify plants and other specimens.

Seaweeds

Fungi

Fungi belong to the vegetable kingdom, but differ from the majority of all other plants by the fact that they do not possess the plant pigment chlorophyll, which gives the green color to plants in general. This green pigment, in the presence of sunlight, enables the plant to manufacture organic compounds necessary to maintain life. Without the pigment the fungi cannot do this and have to obtain their organic food ready made. This they do by utilizing dead plants and animals or by attacking live plants and animals. If they attack dead material they are said to be saprophytic and if they attack living tissues they are called parasitic fungi.

Fungi also lack seeds, reproducing by spores which look like dust or powder on the gills of the fungus. Spores differ from plant seeds in that they have no embryo or food supply.

Some fungi have attained a state of equilibrium with a plant on which they live and these associations are called symbiosis. Lichens are symbiotic plants. They are organisms which are fungi and algae living in close association and differing from either algae or fungi.

These interesting plants are usually found in abundance in early autumn, although some varieties occur throughout the year. They may be collected as rusts growing parasitically on other plants, or as molds growing on jam, bread, paper, fabrics or leather. The dry rot fungus that attacks

timber starts in areas of dampness and spreads by producing its own moisture for further growth. The hard, woody fungi growing on trees and other vantage points need very little attention apart from careful drying. Fungi are often covered with small insects and their larvae, and these should be removed before drying commences.

The fleshy fungi (mushrooms and toadstools) are the kinds likely to interest collectors because of their great variation in shape, form and color. These fungi have different habitats, as do the flowering plants, and each habitat has its characteristic fungi. Some fleshy fungi can be pressed, but most may only be properly preserved in a solution of formaldehyde (usually 5 percent in strength).

The places to search for fungi are damp, moist places, usually found near streams or pools. They are often hidden among long grass or ferns. Therefore, examine the ground in front of you before standing on it.

There are a number of deadly species of fungi. Unhappily, there is no rule-of-thumb method by which to recognize them from the innocuous types. These dangerous types are not even restricted to one genera, but are distributed over a large range. Approximately 90 percent of all deaths from fungi poisoning are due to eating the 'Death Cap' or 'Destroying Angel.'

These fungi show the great variation of types within their group.

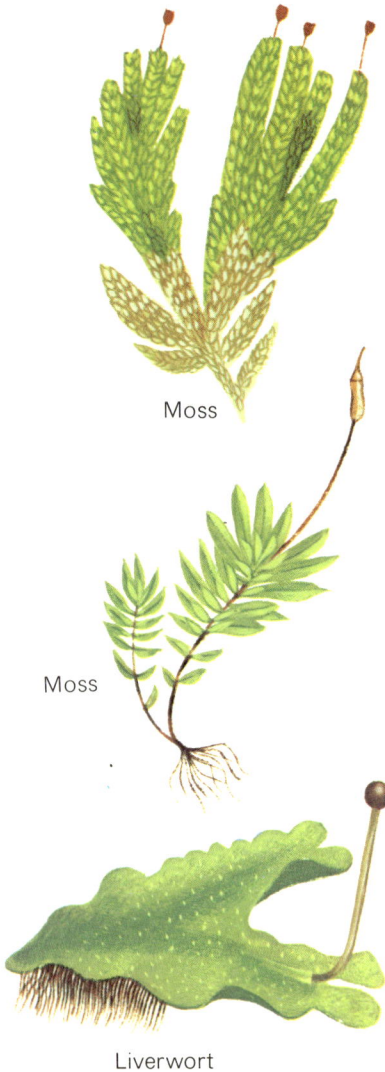

Moss

Moss

Liverwort

Lichens

These plants are really dual organisms, formed by an association of two plants, a fungus and an alga. They are a well distributed group of plants, and occur on walls, trees exposed to high winds, exposed rock surfaces and on the ground. They are one of the few plants to be found in both polar regions, where most plants fail to establish themselves.

When collected, the lichen is allowed to dry in matchboxes or on leaves if it is growing on some. Lichens are kept in small polythene bags with a few grains of naphthalene or paradichlorobenzene to prevent mold growth. Some gelatinous forms require special care with the pressing and drying.

The fungus, usually the most conspicuous portion of the plant, supplies the alga with water and minerals and may prevent it from dying. The alga manufactures carbohydrates which may also be used by the fungus.

Liverworts and mosses are fairly small, but in spite of this are extremely attractive.

An arctic scene, showing reindeer browsing upon reindeer moss (*Cladonia rangiferina*), which is really a grayish-green lichen.

Mosses

Mosses are small, green leafy plants that grow on soil, rocks or bark of trees. Common all over the world, they are most abundant in damp situations such as moist woodlands, but some are actually aquatic, growing submerged in streams while others may grow directly upon stone walls. All mosses lack true roots and stems; their leaves do not have veins as found in higher plants. A branched filament, the protonema, which develops from a spore, differentiates them from the liverworts.

One of the largest mosses, Sphagnum, is also probably most important economically. This moss has exceptional water-holding powers and is often used for packing plant material or germinating seeds. It forms floating mats in small lakes and ditches and may eventually cause development of a bog. Peat is the decayed remains of large concentrations of this bog moss.

Some other mosses rather commonly found are the Fern Moss, Haircap Moss, Broom Moss, White Cushion Moss and Purple Horn-tooth Moss. Although mosses may be allowed to dry in a warm place, it is perhaps more interesting to set up terreria with many different living specimens.

Liverworts

Liverworts are small moss-like plants living in damp conditions, particularly the sides of ditches and ponds. They differ from the mosses in not possessing the protonema, a branched filament which develops from moss spores and from which moss plants develop as lateral buds. Liverworts are much more leaf-like in appearance than the mosses. They can be dried in a plant press, and some of the more fleshy kinds can be preserved in 15 percent formaldehyde solution.

These plants will regain their flexibility if soaked in water, even after many years of drying. Many of the smaller species are very interesting and are often overlooked. They should be looked for on the bark of trees, bare earth, rock surface and other such places. Spray-covered rocks around waterfalls are good places to find specimens of these plants. A few species live entirely in still and running fresh water.

One of the best examples is the Great Scented Liverwort, *Conocephalum conicum,* so called because of its pleasant smell when pressed. This species is distributed on wet rocks and other areas at the margins of streams.

Floating Crystalwort, *Riccia fruitans,* usually found floating just beneath the surface of streams and ponds, can also be found in ditches which are permanently damp or on wet mud where it is most likely to fruit. The shape of the plant varies according to its habitat—on mud it is a thick, violet-colored plant whereas when floating it is slender and very green, with none of the root-like growths present on mud-living varieties. The Purple-fringed Riccia, which grows in wet areas, has slim purplish scales under a deeply furrowed, floating thallus.

One of the largest leafy liverworts, Three-lobed Bazzania, forms mats in cool places. Its upright branches have three-toothed leaf tips and coarsely toothed underleaves.

Common Liverwort has a much-branched, ribbon-like thallus with internal air chambers forming a diamond-shaped pattern on the outer surfaces. Separate male and female plants each produce umbrella-like sex structures. This species is often so successful that it is considered a weed.

About 300 species of hornworts are classified along with the 8,500 species of liverworts.

3-lobed Bazzania, a liverwort

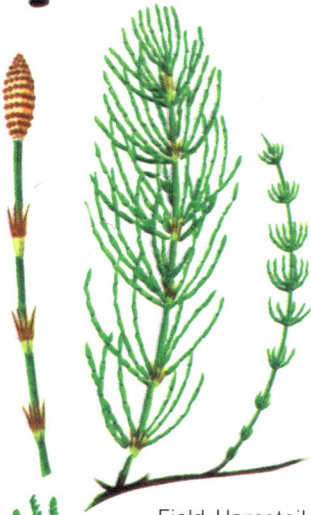
Field Horsetail

Running Ground Pine

Fern Allies

Remnants of ancient groups of plants are whisk broom ferns, horsetails, quillworts, club mosses and spike mosses. With little commercial value, these plants are important as ground covers, holding moisture and sheltering smaller plants.

Horsetails have ridged, segmented stems containing a lot of silica. One of 25 species in North America, Rough Horsetails or Scouring Rushes, have so much silica that they were used by colonists for scouring pots and pans. Most common is the Field Horsetail. It can be found in a meadow or field, by a railroad or by the edge of a pond or lake.

Most club mosses have small, scale-like, evergreen leaves arranged in a whorl around an erect or creeping stem. Many have been thoughtlessly gathered for Christmas decoration and florist use. Fortunately there still are areas where a large number of plants can be seen blanketing forest floors. Some species are Princess Pine and Ground Pine.

Quillworts are mostly found in wet areas. Their bulb-like base is important as waterfowl food. Best known of the spike mosses is the Resurrection Plant.

Bracken growing under the diffused light filtered by trees

Ferns

Ferns also reproduce by spores but show evidence of their higher development by having true roots, stems and veined leaves or fronds. The spores develop into minute liverwort-like prothallia from which tiny fronds uncurl.

Although most numerous in the tropics, ferns are plentiful in temperature regions from wet areas and shaded forests to rocky cliffs and open meadows. In all, there are 250 native species north of Mexico as well as some introduced species.

World-wide in distribution, Bracken forms dense growths in poorly soiled fields and woodlands. In some areas, its many-branched fronds may be 6 to 8 feet in height making it one of our largest ferns. It can be found from coast to coast, Canada to Mexico.

Cinnamon Ferns are common in wet areas. Their tightly coiled crosiers, or 'fiddleheads,' have thick wooly coatings when they push up in early spring. Separate spore bearing fronds appear in spring before the sterile leaves. As the spores ripen, they turn a rich cinnamon color as do the silvery hairs on crosiers and bases.

Hay-Scented or Boulder Ferns are so common in pastures and dry woodland that they are often considered weeds. They have a lacy appearance and sweet fragrance which has given them their name.

Found in damp places from the Gulf States to Canada is the Sensitive Fern which wilts quickly when picked and is one of the first ferns to succumb to frost. Its spores are encased in hardened, beadlike leaflets on separate, stalk-like fertile fronds.

The Ostrich Fern which grows from Virginia to the far north has fertile frond leaflets which curl up forming a 'pod' around the spores. Its sterile leaves are large and plume-shaped—wide near the top and tapering to the bottom.

Boston Fern is most likely to be potted as a house plant, although it grows wild in southern Florida. Its 'Boston' name results from the fact that a number of varieties were developed near that city. It is of commercial importance.

Some put great emphasis on the role of ferns in the production of coal, but they are probably more important as soil builders. Several ferns, such as the Common Polypody and Spleenworts, grow on top of rocks where there is little soil. Their decaying parts add humus in which larger and more advanced plants may grow.

Sensitive Fern   Cinnamon Fern

Larch

Scots Pine

Juniper

The Seed Plants

This group of plants is called the spermatophyta and is usually divided into two groups: the gymnosperms, including the conifers with naked seeds, and the angiosperms, which include the remainder of the flowering plants with enclosed seeds.

Gymnosperms

Gymnosperms include the pines, larches and cedars, among many others. They are found in all parts of the world except the tropics, and even there they may occur on high ground. The fruits of these plants are called cones, and these contain the naked seeds. Cones are interesting objects to collect; they vary greatly in shape and color, and may be sprayed with dilute glue or adhesives to prevent the seeds from falling out. Some examples of this group of plants, the yews, have a fleshy fruit called an aril, which is bright red when ripe. These fruits should be preserved in fluid.

Gymnosperms are either trees or shrubs and all show great variety. They are found from the earliest times in fossil form, being found far back in the Paleozoic period.

They appeared to be tall trees with woody branching stems and simple leaves.

Plants living today which resemble these ancient groups are the cycads and the gingko, or Maidenhair Tree. This tree used to be found only in the precincts of temples in China and Japan, although it is believed that it grows wild in Western China. Very fine examples are seen in our botanical gardens, and small trees are now available for sale by horticultural suppliers.

One of the most important groups of living gymnosperms are the conifers. These are evergreen trees and shrubs, with greatly reduced leaves which are usually needle-shaped. These plants reach the limits of vegetation in the areas around the polar regions and in mountainous districts.

There are many varieties of conifers to be seen and these include larch, cypress, spruce, juniper and pines.

The Sequoia groves of inland California and Redwood forests of the West Coast are interesting and beautiful. Redwoods of a 350-foot height and Sequoias with 35-foot diameters are impressive sights. Surely these deserve our protection.

Spruce

Yew

Aril of Yew

Yew—
female flowers

Angiosperms

Angiosperms include the rest of the flowering plants, and are classified into two main groups: monocotyledons and dicotyledons, usually shortened to monocots and dicots. The bulk of an angiosperm seed is made up of one or two leaf-like structures called cotyledons that are packed with food, especially starch, with which the young plant begins its development. Monocots have one of these leaf-like structures and the dicots have two such organs.

In monocots, the leaves of the plants have parallel veining, whereas in the dicots, the veins are branching and net-like in appearance. Monocots include the grasses, lilies, orchids,

Lady Slipper

Iris

Trout Lily

Ragged Robin

daffodils and some very large plants such as the date palm. Dicots include such plants as sunflowers, dandelions, buttercups, potatoes, shrubs and trees.

The monocots include the important food plants, such as the cereals and fodder grasses. Their vascular systems are in the form of closed, scattered vascular bundles, and the flower parts are usually in threes or multiples of three.

The dicots have a ring of vascular bundles. The flower parts are in fours or fives or multiples of these.

Corn, by the way, is a monocot, as it belongs to the grass family. Many species of this native plant have been developed since it has been cultivated.

Daisy

Ground Ivy

Dandelion

Dog Violet

General Preservation of Plants

As previously described on page 12, the plant press can be used for the majority of specimens collected. Fungi and other fleshy plants can be dried in an improvised sand bath and some flowers can also be treated in this way. A tin with a depth of 4 to 5 inches is used and a layer of washed sand poured onto the bottom. The plant to be dried is placed on the surface of this layer and arranged in the position required for the collection. Meanwhile another sample of sand is being heated. When this sand is quite hot, usually after about 15 minutes of heating, pour it over the arranged plant and leave the sand bath to cool. This is an effective way to preserve many plants as they are found in nature and is sometimes preferred to the flattening of the plant press. In addition, the flowers are sometimes well preserved by this method.

After pressing or drying, the plants are usually attached by means of gummed canvas strips to sheets of good quality white paper and kept in folders or files. The plants may also be carefully glued in place on the paper. The container in which they are kept should be airtight, and a few crystals of

When pressing plants having thick stems, pad out the spaces between them with paper.

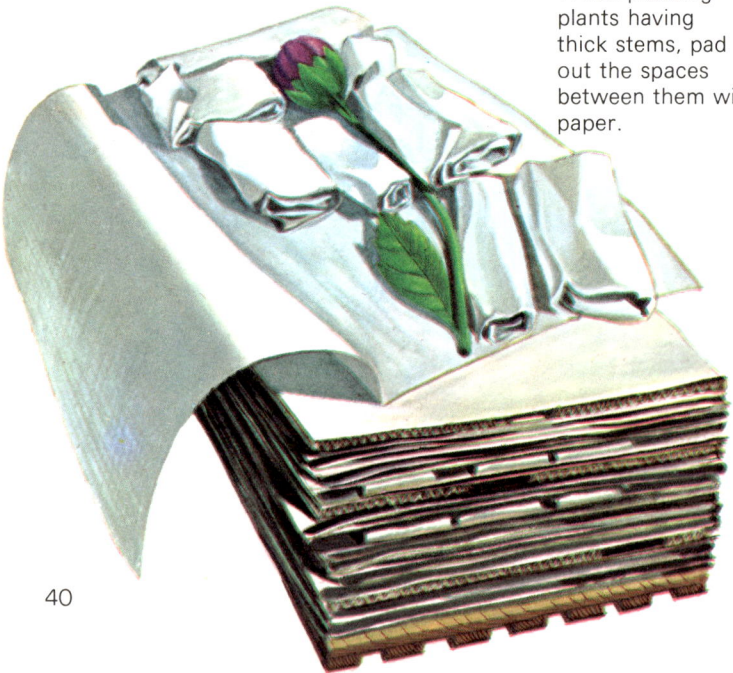

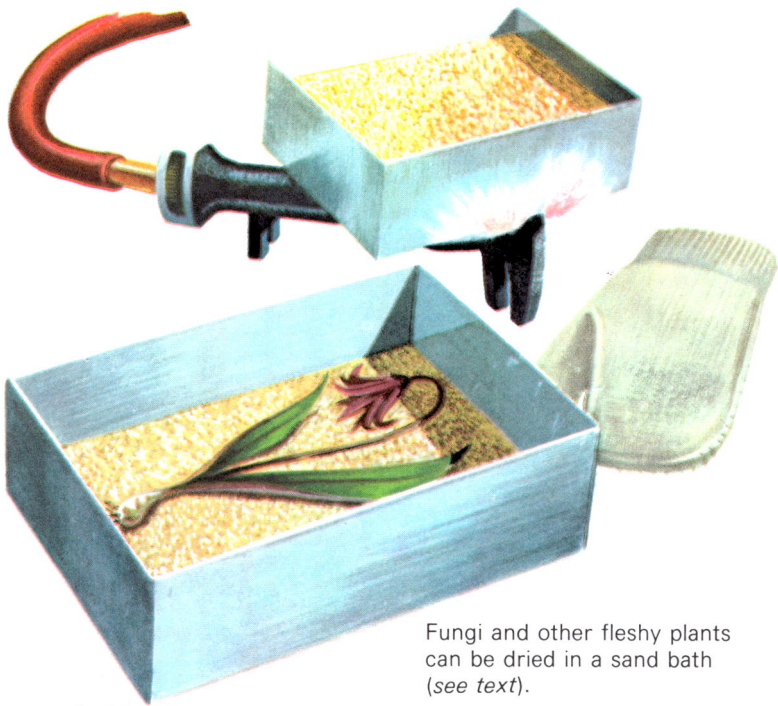

Fungi and other fleshy plants
can be dried in a sand bath
(*see text*).

paradichlorobenzene should be put in the container to keep out insect pests and molds. These collections of pressed plants are called herbariums.

Special labels can be purchased to stick on to herbarium sheets by which the 'pressed' plants can be identified. These labels have spaces set aside for common names, scientific names, family and area in which the plant was found.

Because the labeling of these pressed plants often takes place a month or more after the plants were found and placed in a press, it is best to write as much detail as you can in a small pocket notebook when the plant is collected. Trying to remember details of where the plant was found several weeks after collection may be impossible to do.

Many plants will have to have their roots washed before they are pressed. Dry them as much as possible before setting them in the press. Although it is recommended that tough, thick plants are best left to wilt before being pressed, it is then extremely difficult to set the leaves in position if left too long, especially if they begin to shrivel.

Spore Prints

Spore prints made from fungi are a very interesting way of recording the various types of fungi collected, especially as fungi do not keep very well in collections. The cap of the fungus is laid, gills downward, on a sheet of white paper and left overnight. Spore powder is deposited and the print can be sprayed with varnish to preserve it as a permanent record. Leaf fungi are best preserved by pressing.

Preserving Green Color in Plants

Occasionally it is interesting to try to preserve the green color in plants. This is done by boiling the collected specimens in a dilute solution of vinegar and copper acetate.

Fungi are often best preserved in fluid, to ensure that they are not squashed.

The copper acetate is added to the vinegar until no more will dissolve. The mixture is then diluted by half with water and the solution used to simmer the plants. It is usual to keep a fresh specimen aside to see how the progress of color restoration proceeds. At first the green plant turns brown, and then the green color slowly develops until the two plants, the living specimen and that being preserved, look alike. The preserved plant is then washed under the tap and allowed to dry naturally in air or under the press if required.

Photography

Photography is yet another method by which a record of plants and animals can be kept. It is an exceptionally good way to record the colors of plants.

Freeze Drying Animals and Plants

One of the most recent methods devised to keep both plant and animal tissue in a good state of preservation is by 'freeze drying'. Briefly, this technique consists of freezing the animal or plant and then subliming away the water in the ice crystals under pressure so that the specimen dries in an unchanged shape.

Vinegar and copper acetate can be used to preserve the green color of plants. Specimens can be compared with freshly picked plants.

This method, of course, has the advantage over taxidermy or 'pressing' plants by not damaging the specimen.

Notebooks

Just as important as any colorful pictures or collection of plants or animals is a good, neat and well kept notebook. Not only will such a book enable specimens to be readily identified, even after a few weeks, but it will also help to create a scientific record and attitude.

COLLECTING ANIMALS

Within the large classification of invertebrate animals there are many well-defined sections. These sections are listed in the table set below. These creatures range from the protozoans to the chordates, and include insects, eelworms, snails, spiders, centipedes, millipedes, sponges, jellyfish and corals.

Many of these creatures are easily studied within their own environment, so providing the observer with a fascinating hobby.

Protozoa	Small single-celled animals
Porifera	The sponges
Coelenterata and Ctenophora now collectively called the Cnidaria	The jellyfish, corals, anemones and sea gooseberries
Platyhelminthes or flatworms	Planarians, tapeworms and flukes
Nemathelminthes or roundworms	Eelworms, *Ascaris* and vinegar eels
Rotifera	'Wheel' animalcules
Bryozoa	'Moss' animals
Annelida	Polychaeta, Leeches and Earthworms
Mollusca	Snails, mussels, slugs, squids and octopus
Arthropoda	Insects, crustaceans, spiders and centipedes
Echinodermata	Starfish, sea urchins, sea cucumbers and sea lilies
Hemichordata	Acorn worms
Chordata	Sea squirts and vertebrates

The Invertebrates
Protozoans (Protozoa)

These are minute single-celled animals, found in ponds, ditches and streams and also in the sea; a few are found in moist areas on land. There are some which are parasitic, a famous example being the malaria parasite, which attacks the blood corpuscles of man, many animals and birds. Many

Protozoan
(Ciliate)

Poriferan
(Sponge)

Cnidarians
(Hydras)

Flatworms
(Planarians)

Mollusk
(Snail)

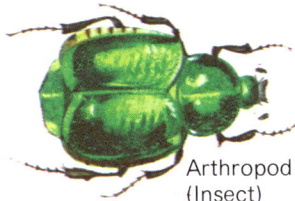

Arthropod
(Insect)

Arthropod
(Spider)

Rotiferan

Bryozoan

Annelids
(Sea Worms)

Nemathelminthes
(Roundworm)

Chordate
(Sea Squirt)

Echinoderm
(Starfish)

Hemichordate
(Acorn Worm)

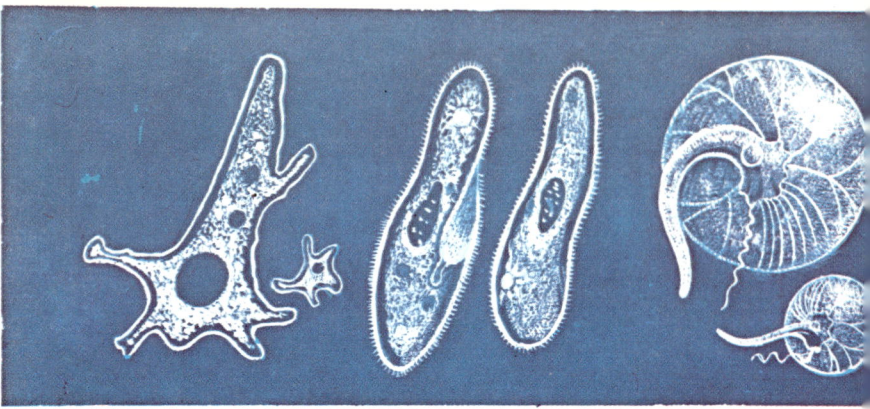

A wide variety of forms of protozoa, ranging from the fresh-water types at left (*Amoeba* and *Paramecium*) to the marine types (*Nociluca* and *Foraminifera*) at right

of the marine forms secrete shells, which after the animals' death form huge deposits on the sea-bottom. Chalk cliffs and quarries are composed of countless shells of long dead protozoans. The pyramids of Egypt were built from blocks of a limestone that consisted almost entirely of large shelled protozoans called nummulites.

Protozoa are able to encyst, enabling them to be carried about by other animals and to live in areas otherwise not available to them. The cyst survives temperature extremes and drying. After rain or transfer to a wet habitat, the cyst will break open and the protozoa are liberated to move about once again.

The protozoan is very much like a single cell, except that the functions of many groups of other cells which together form tissues and organs seem to operate in the single-celled animal.

As previously mentioned, many are parasitic and have a role which seems to be parasitic, but the animal which houses these creatures seems to suffer no harm. Opalina found in the rectum of many amphibians and easily visible to the naked eye is one of these, and so too is monocystis found in the seminal vesicles of the earthworm, again seemingly causing no harm or damage to its host.

Some protozoans are necessary for good health, such as Trichonympha found in the gut of wood-eating termites. For a long time it was a mystery how these termites could exist on a diet of wood, since wood contains only minute amounts of protein and sugar.

It has been found that these protozoans ingest minute particles of wood and transform them into soluble utilizable food for the termite.

A few protozoa, however, are killers. They cause serious diseases in both man and animals, such as African sleeping sickness and malaria. Malaria, caused by a plasmodium, is carried by the Anopheles mosquito. Injected into a human together with the mosquito's saliva, the amoeba-like sperozoites enter the red blood cells. Each of these then forms spores, which multiply and break out at regular intervals, causing chills and fever.

Some spores develop into sexual forms, which the mosquito takes when it bites an infected person. In this way, the disease is quickly spread from person to person.

In a mosquito's stomach these sexual forms form eggs and sperms. Each fertilized egg develops in a capsule, in which many new sporozoites form. These migrate to the salivary glands, and are then injected as the mosquito bites.

Nummulites, which are found in certain rocks from which the Pyramids were built

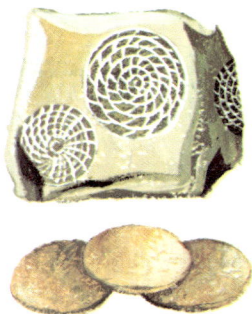

Marine Sponges

Sponges (Porifera)

This great group of animals is widespread, occurring both in fresh water and in the sea. They are classified according to the structure of their supporting tissues. These may consist of a spongy, horn-like substance, or of skeleton-like spicules composed of chalk or silica. Examples of these creatures include the familiar yellow bath sponge from the West Indies and the Mediterranean, the brilliant green fresh-water sponge found encrusted on lock gates on slow-flowing canals and rivers, and the exotic Venus Flower-basket found in the deep sea, which is composed of intricate glass spines.

Sponges, particularly the fresh-water varieties, have astonishing powers of regeneration; if a sponge is pressed through a fine sieve, the resulting soft green mass will resolve itself slowly again into the familiar sponge shape. A number of interesting examples of sponges inhabit the seashore. Among these is one that bores through mollusk shells, especially oysters, and even limestone rocks are often found with bore holes in them. Some sponges are brightly colored—yellow, orange, red and even blue.

Collectors can dry out sponges quite easily after carefully washing them well in fresh water to remove the debris and mud that they accumulate within themselves.

Cnidaria

The cnidarians, which together with the ctenophores, were formerly called the coelenterates, include the jellyfish, sea anemones and corals.   The now separated ctenophores are represented by the sea gooseberries. Coelenterates are all aquatic and most live in the sea. Very few live in fresh water. The hydra is a common example of a fresh-water type.

The body of a coelenterate is built on a simple plan. It is more or less radially symmetrical, with a jelly-like consistency. These creatures have guts with only one opening. There is a nervous system although it is hard to demonstrate its presence. There is no excretory system or any evidence of a blood system. The body wall is diploblastic —having two layers. Between these layers is the mesoglea — a jelly-like material. In hydras, this is very thin while in anemones and jellyfish it is tough and fibrous.

Hydroid colonies, common in lower rock pools along the seashores, may be collected in spring. The fascinating liberation of the medusa form from the branching reproductive tentacles can be watched under a hand-lens. These tiny jellyfish are part of the life history of these colonial coelenterates and proof of the relationship between animals in this group.

Hydroid colonies

Jellyfish of many kinds are found swimming in the open sea, but many are caught by the tide and may be found in rock pools or on the shore after the tide has retreated, forming shapeless and often large jelly-like masses.

The vivid blue gas-filled float of the Portuguese Man-of-War is a familiar sight in warm seas. Occasionally, when conditions are right, these jellyfish 'invade' temperate waters and many are washed ashore from New Jersey to Florida during these times. The animal is really a complex floating colony and may reach a length of over 50 feet with its masses of long tentacles. These tentacles can inflict very painful injuries, which are occasionally fatal. The injuries are caused by stinging cells called nematocysts, which are used by the animal in capturing its prey. A similar, although smaller animal than the Portuguese Man-of-War, is the Velella or 'by-the-wind sailor'. This animal also has a gas-filled float but is only some 2 to 4 inches in diameter. It is an inhabitant of the Gulf Stream but is sometimes blown ashore from Maine to Florida.

Besides the Portuguese Man-of-War, many other colorful and interesting jellyfish may be seen in coastal waters. The Moon Jellyfish or White Sea Jelly *(Aurelia aurita)* is washed up on all our beaches. Its milky disc-shaped body has pink and white sex organs forming a four-leaf clover pattern. The body usually measures 3 to 10 inches across.

Portuguese Man-of-War

The Pink (or Red) Jelly-fish *(Cyanea capillata)* is the giant jellyfish of the At-lantic. Although many along the New England coast are about a foot in diameter, rare specimens reach a size of 8 feet. Fortunately, the largest specimens are in the coldest waters and since it ranges northward to Arctic waters, the fatal stinging power of its 800 tentacles is not too often inflicted upon swimmers.

Many sea anemones may be found in rocky pools; they can resist the tidal conditions better than many of the shore animals and shrink into a small ball until the return of the tide when they regain their familiar flower-like shape.

Most common is the Brown Anemone. Four inch-es high, with a stalk over 2 inches across, it is also the lar-gest of the North American anémones. Large specimens may have up to a thousand waving yellowish tentacles.

The small fresh-water polyp, called Hydra, is abun-dant in still, clear ponds and brooks. It is found clinging to the underside of duckweed or to submerged leaves of aquat-ic plants. To preserve these interesting creatures, it is

The life history of jellyfish— from the adult stages at top, clockwise through the planula, hydrula, strobila and then into living ephyra stages, swimming off as tiny jellyfish

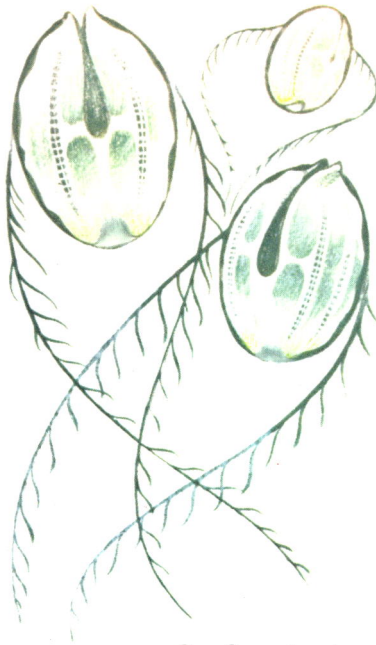

Sea Gooseberries

Anemones

necessary to first quiet or relax them. This is done by placing one or two of the animals in a small dish with a little of the water that they were taken from, and adding a crystal or two of menthol to the surface of the water. In a few minutes, the hydra will cease to move and will not respond to a touch. If this is not done, the animals will contract to a shapeless mass of cells. They may then be preserved in a 70 percent alcohol or 5 percent formaldehyde solution. Sea anemones and corals can be preserved in the same way, but it is best when dealing with marine creatures to make up a preserving solution with seawater instead of fresh water. Larger jellyfish can be preserved in a 10 percent formaldehyde solution for a few days and then transferred to a 5 percent one for storage.

Fresh-water hydra may be kept alive in small ponds and aquariums and should be fed small water fleas called daphnia or other small crustaceans such as cyclops. Sea anemones may be kept in marine aquariums and fed pieces of shrimp or mussel. Small pieces of meat and liver are also acceptable. Care must be taken not to overfeed these animals or the excess food will cause contamination and poisoning of the aquarium very quickly. The ctenophores, or sea gooseberries, can be found drifting off-shore, often in great numbers, and are sometimes left behind by the tide in rock pools. They are difficult to preserve and do not live long in an aquarium. It is best to observe them without capturing them.

The tube anemones are long and slender, and live buried in sand almost to their feeding discs. The slender tentacles arise in two distinct sets—an inner and smaller set encircling the mouth, and a marginal set, each rounded by a tube formed of a hardened slimy secretion.

Corals are extremely interesting creatures to collect and observe. Unfortunately, most really large corals are found in tropical seas. Huge reef colonies exist in which millions and millions of individuals live. Many corals can be found in more southerly waters.

One of the few stony corals found from Cape Cod south to Florida is the Star Coral. It only thrives in shallow, unpolluted water but broken skeletons may often be found washed up on the beaches.

Sea Fans, Sea Pens and Whip Corals are 'horny' corals. These have a flexible, rather than stiff, skeletal core of a horny material called gorgonin. They are colored varied shades of red, orange, yellow and purple. Due to their delicate structure—complete skeletons are not often found washed on shore—they must be collected as living specimens.

Corals

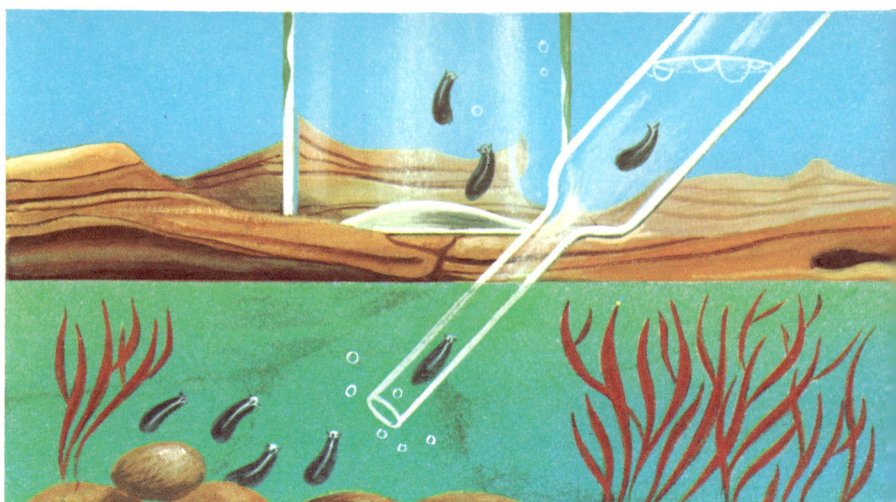

Collection of water animals being sucked up a tube and placed in a jar

Flatworms (Platyhelminthes)

Free-living flatworms are found in the sea and in fresh water and may also be found in damp places on land. They may also live in mutual habitats with other animals, and some flatworms are parasitic.

One of the commonest free-living flatworms is the planarian. These creatures should be searched for under stones and among plants in fresh water, in rock pools on the seashore, and under fallen logs in damp places in woodland areas. Sweeping water weeds with a fine-meshed net is sometimes a successful way of catching them. These animals may be quieted by placing a crystal or two of menthol in a small tube containing the worms, and they may then be preserved in 70 percent alcohol or 5 percent formaldehyde solution. Some of the larger leaf-like forms may be flattened during their preservation between two small pieces of glass (microscope slides are excellent for this purpose).

The leaf worms or flukes are a very varied group, occurring in many animals and having a complicated life history. Many kinds pass their early life stages as free-swimming larvae and then enter a host animal, sometimes a fresh-water snail, before passing into a more advanced stage when they encyst on grass. The grass is eaten by sheep, then the adult fluke develops in the liver and lays eggs. The cycle starts again; the eggs pass out of the host and enter the water.

Tapeworm

Echinococcus

Gyrodactylus

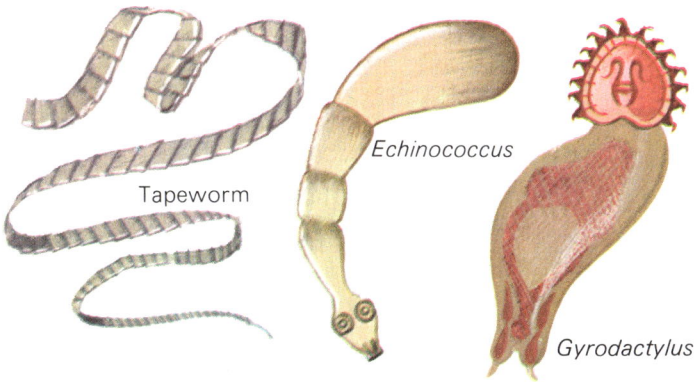

Flukes are normally obtained in a slaughterhouse from fluky sheep or cattle, and are usually dissected out of the bile ducts and major blood vessels. The worms are placed in a solution of salt and water (about 10 percent) to wash off debris, and they are then fixed and preserved in the same way as the planarians. Tapeworms and flukes are often found when examining frogs and other animals dissected in schools.

Ribbon Worms (Nemertinea)

These worms are found in rock pools on the seashore and in damp places on land. They are able to glide along smoothly, but they can be quieted in dilute (2 percent) formaldehyde and then preserved in a 5 percent solution.

They are able to extend to an extremely long length, and then contract to a fraction of that length.

Life cycle of Chinese Liver Fluke

Life-cycle of Beef Tapeworm

Ribbon Worm

Roundworms (Nemathelminthes)

These are white worms, which retain their shape at all times without contracting. Most roundworms are free-living in nature, but a number are parasitic on plants and animals. The larger parasitic varieties are found in horses and pigs and other similar animals. Man is occasionally infected from these sources. The smaller examples are found in the sea, fresh water and soil. Some cause disease in potatoes. There is one variety that lives its entire life in malt vinegar, the so-called Vinegar Eel.

The worms are usually pointed at both ends with no sign of the segmentation seen in the higher groups of worm-like creatures. There is a cuticle which is shed frequently. There is usually a straight gut, with a mouth armed with hooks, and a simple anus. The female has a separate opening for eggs, although in the male the sperms pass out of the anal aperture. There are no cilia, no evidence of a blood system and no organs of respiration. The eggs are usually covered

Roundworm in liver

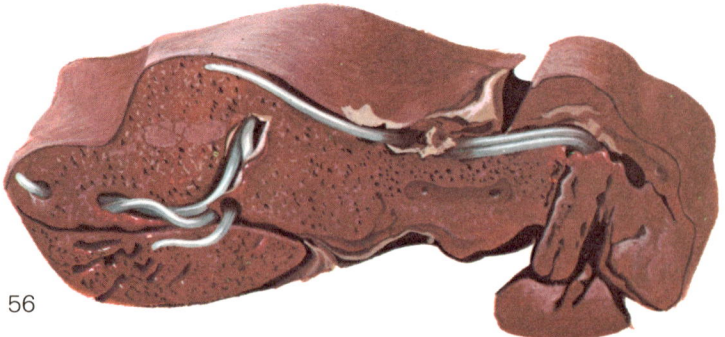

with a hard shell. Sometimes, the eggs develop inside the female and can be seen dividing and growing through the transparent wall of the egg. The eggs may encyst, enabling the young worm to resist drying and other hazards.

Many of the roundworms are parasitic for part of their life. Many attack plants. They move by a series of S-shaped curves. This threshing movement does not result in much forward movement in water, but when in earth the worm can get along quite easily.

One of the largest of the nematodes is *Ascaris,* a species of which infects man. The adult worm sometimes measures up to a foot in length. The males are smaller than the females and have a curved posterior end. Approximately 200,000

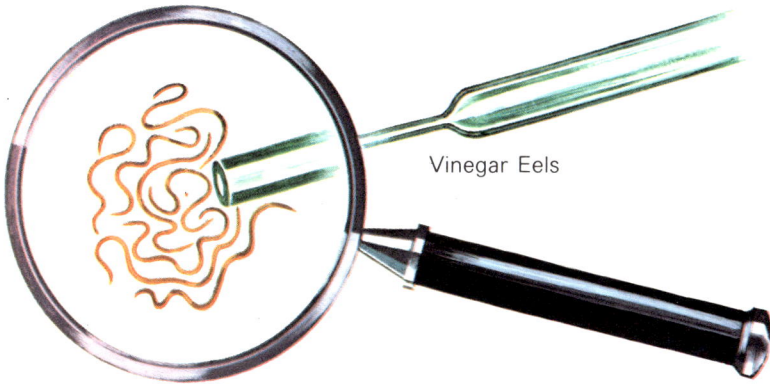

Vinegar Eels

eggs can be laid daily by the female worm. These eggs are passed out in feces onto the ground and there develop into little worms. If the eggs are laid on vegetation, they may be ingested into the intestine of man after eating and will grow there. They do not remain there long, soon starting to migrate from the intestine into the blood vessels, through various organs and finally pass into the lungs. From the lungs they move up the bronchii to be swallowed again and passed down into the intestine. Here they remain, growing rapidly. They resist digestion by secreting a substance that counteracts the action of the hosts' digestive juices. The greatest damage is done during these migrations.

The adult worms in the intestine seem to be fairly harmless.

The female hairworm lays a string of eggs around water plant stems.

The hairworms resemble the roundworms and are often classified in this group. Gordius is a common kind which wriggles around in ponds and ditches. The name of these creatures, gordian worms, is taken from the tangled masses, or gordian knots, in which they are often found. They are unusual in that the young stages are normally parasitic on insects, while the adults are free-living in fresh water. They often appear suddenly in rainwater butts or horse troughs. The female lays long strings of eggs wound around water-plant stems. The hatching larvae bore their way into aquatic insect larvae and develop into adults as their insect hosts grow, or they may escape and transfer themselves to another insect host, finally escaping from this insect to mate and lay eggs again in the water.

These animals are usually preserved in 5 percent formaldehyde solution, being washed in salt water to remove any debris and other matter.

Rotifers

The rotifers, or 'wheel' animalcules, are small creatures just visible to the naked eye. They are so-called because of the crown of rotating hair-like processes at the head end. Rotifers live in fresh water and in damp moss, and are found less commonly in the sea. They are usually preserved when other animals are collected, as they are often tentatively attached to the other animals.

Rotifers can resist drying even more than protozoans and nematodes. In this almost completely dry state they may remain alive for years. As soon as moisture is present they reactivate and swim about rapidly. Because of this resistance to dry conditions these creatures can live in situations that are only damp on few occasions, such as roof gutters and rock crevices, as well as among mosses. When the water disappears, the rotifer shrinks to very small volume and loses most of its water content. Often the animal dies, but the eggs contained inside survive and when moisture returns will become active. There are a few marine rotifers, but the greater majority live in fresh-water and damp fresh-water areas. Because of their resistance to drying, these animals are distributed over large areas.

Some rotifers can be easily seen with the naked eye. The tubes that they secrete to live in can also be seen. *Floscularia*, formerly called *Melicerta*, was sometimes called the brick builder because of its habit of forming a cast of small pellets made from the surrounding mud. Some of the early microscopists used to provide this animal with colored particles so that it would produce multicolored tubes.

Two other rotifers are worth mentioning. The tube dweller *Collotheca* has a gelatine-like tube, inside which eggs often can be seen. The nicest of these creatures is no doubt *Stephanocerus* which, like the previous animal, lives in a gelatinous tube. It is a large species, growing to over 1 mm in length. Under a hand-lens it is a beautiful sight.

A collection of rotifers

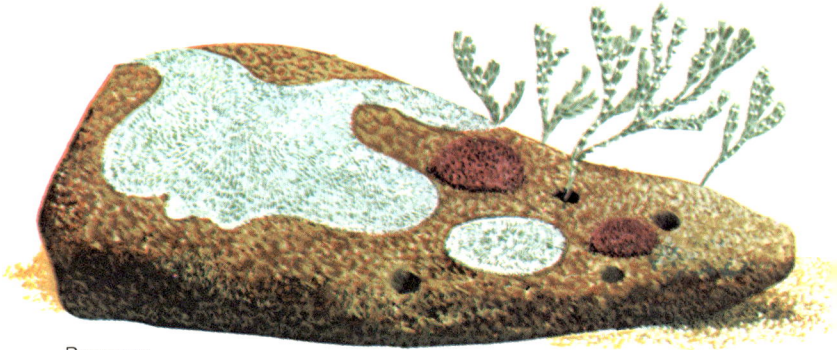
Bryozoa

Bryozoa

The bryozoa, or so-called zoophytes or moss animals, form branching colonies in the sea called sea mats or lace coralline; they are also found attached to rocks, stones and seaweeds. The fresh-water forms are generally more delicate in shape and form gelatinous masses on submerged stones, roots of trees, floating plants and so on. They can be dried out quite successfully. The fresh-water gelatinous forms can be preserved in 5 percent formaldehyde.

Segmented Worms (Annelida)

These creatures, readily identified by the regular ring-like segments dividing the body from one end to the other, are found in the sea, on land and in fresh water. A few are parasites in other animals.

The segmented worms are grouped into three classes: the sea worms, or Polychaeta; the earthworms, or Oligochaeta; and the leeches, or Hirudinea.

The Polychaeta

Most of these worms are marine and are usually found by their casts on the seashore after the tide has retreated. Small kinds may be found by looking into rock pools or by turning over stones and rocks. Digging often reveals many interesting kinds. The fisherman will be familiar with the clamworms and bloodworms, found by digging on the shore and used for bait in fishing. Many of these worms live in chambers that they build in the mud, gravel or sand, and many have long tentacles, giving them the appearance of exotic flowers. They have great powers of retraction, and the flowers will disappear in an instant if danger threatens.

Most of the worms found on the shore can be quieted by adding some Epsom salt to the sea-water, and a few menthol crystals act in the same way. The specimens may then be preserved in 5 percent formaldehyde solution.

The Oligochaeta

Despite their common name of earthworms, many of these worms are found in fresh water, and a few are found on the seashore. They nearly always occur without any obvious feelers or appendages of any sort, which readily distinguishes them from any of the polychaeta. The earthworms of gardens and fields are the easiest to collect. After a spring rain, it is common to see many worms stranded on the pavement, having been forced by the rainwater to leave their burrows. Always remember to put them back on the earth, because otherwise they will dry out and die very quickly. If worms are to be collected, the easiest way to bring them to the surface of the ground is to sprinkle the area with a dilute solution of mustard or potassium permanganate. The worms will come to the surface shortly after the area has been treated. They may then be relaxed in weak alcohol (30 percent), if available, or relaxed in dilute formaldehyde (2 percent) and then transferred to 70 percent alcohol or 5 percent formaldehyde.

Polychaeta

Amphitrite gracilis

Arenicola marina

The Hirudinea

The leeches are flattened segmented worms with a sucker at each end of the body. They are familiar to anyone who goes pond dipping, as they are numerous in fresh-water ponds, ditches and streams. Many are found in the sea, but a few live on land and may become parasites on animals, including man. Among the land leeches, there are some, found in southeast Asian forests, which drop onto men and animals in forests in Burma and other tropical areas and eventually find their way to the back of the throat, where they cling and suck blood. These leeches have powerful jaws and secrete

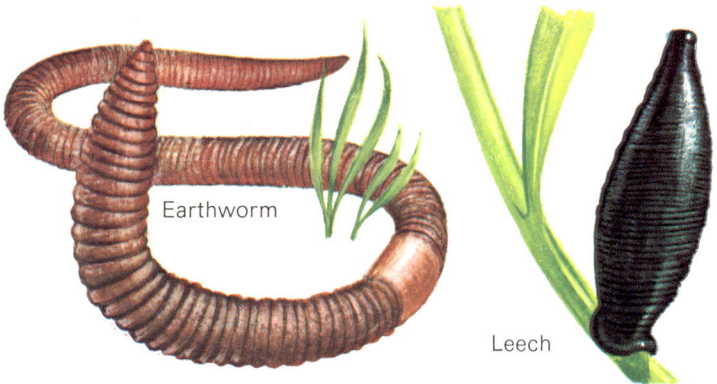

Earthworm

Leech

a substance that causes the blood to continue flowing long after clotting should have taken place. This secretion called heparin is used in medicine. Leeches may be prepared and preserved in a similar way to the polychaeta worms.

Mollusks (Mollusca)

The mollusks are soft-bodied animals. Most people picture clams or oysters when they think of mollusks, but there are others—the octopus, squid and garden snail are among these. The great majority of mollusks have a calcareous shell which is composed of either one or two valves. In some the shell is concealed within the tissues, and in a few there is no shell at all. The coat-of-mail shells have strange shells composed of series of eight dorsal or upper plates. Gastropod (single-shelled) mollusks may be marine or fresh-water creatures, or they may live on dry land. The marine forms are found between the tide levels and at all depths, while a few forms float in the upper layers of the sea. Bivalves (such as oysters, cockles and mussels) are usually found living in mud and sand or attached to rocks. Coat-of-mail shells, tusk shells, and squids, cuttlefish and octopuses are all marine. They live at various depths, either lying on the bottom or swimming freely in the water.

Limpets

Coat-of-Mail Shells (Amphineura)

Coat-of-mail shells, also called chitons, may be collected on rocks and in seashore pools. They can be preserved by attaching them to flat sticks with tape or string and placing them in 5 percent formaldehyde or 70 percent alcohol, or they can be allowed to dry out naturally.

Snails and Slugs (Gastropoda)

These mollusks are easy to find, especially after rain. They have great variety in color and shape. The shells are often used as collectors' pieces, but it is useful to know how to preserve the entire animal. The preservation of these animals depends on good relaxation, and this is achieved by several

Garden Snail

Garden Slug

methods. The land forms (such as garden snails and slugs) can be placed in a container of boiled water that has been allowed to cool. The container should have a closely fitted lid. The mollusks soon use up all the available oxygen and become extended in a state of asphyxiation. Some forms, especially those found on the seashore, will not react so well to this treatment and they must be placed in a container of sea-water with a few crystals of Epsom salt or menthol. This treatment is good for the periwinkle and the small whelks. Limpets are pried off the rocks, placed in dilute formaldehyde or alcohol, and the strength of the solution is slowly increased. If formaldehyde is used, start with 1 percent and work up to 5 percent. If alcohol is used, start at 15 percent and work up to 70 percent.

Bivalve Mollusks
(Pelecypoda)
These animals, such as mussels and oysters, can be preserved by forcing open one valve or shell and allowing a stream of hot, but not boiling, water to enter. A few minutes is sufficient to relax the mollusk. They are preserved in 5 percent formaldehyde solution.

Octopuses and Squids
(Cephalopoda)

Since these animals die very shortly after leaving the sea (they are exclusively marine), the only precaution is to have adequate preservative ready. A solution of 5 percent formaldehyde is used for this purpose. Cuttlebones and squid pens, the internal shells of these mollusks, are sometimes found dried out on the seashore. One of the most fascinating of the molluscan groups are the giant squids. These creatures are far more common than is generally realized. They usually frequent great depths and give rise to many so-called 'sea monster' stories.

The discoverer of these animals described one with a 10-foot-long body plus 42-foot-long tentacles found off the banks of Newfoundland. Surely this could not be preserved, but smaller examples washed ashore might be.

Both squid and octopus swim by jet propulsion. An octopus can also move along using its tentacles which have many suction discs. These creatures change color rapidly when excited and are able to emit an inky fluid which aids in protection as a smoke screen would.

Univalve mollusk (above)
Bivalve mollusks (below)

Octopus

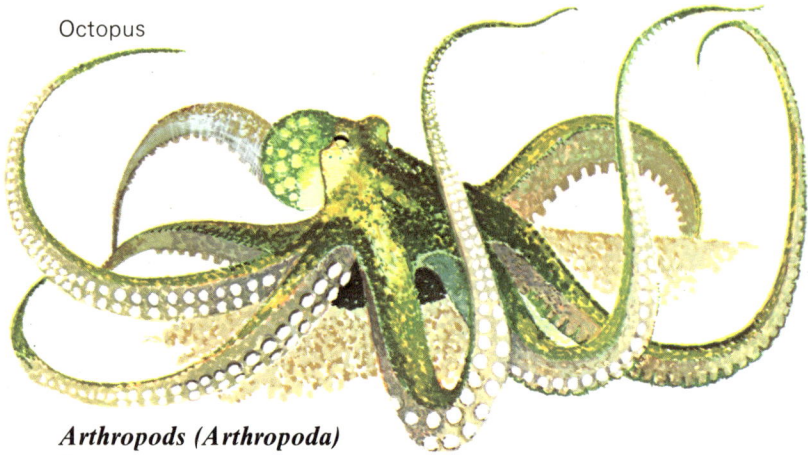

Arthropods (Arthropoda)
Insects

Insects make up three-quarters of this group of animals, which are characterized by the possession of an external skeleton and jointed legs. Their classification is:

Orthoptera	Cockroaches, Grasshoppers, Crickets
Plecoptera	Stoneflies
Dermaptera	Earwigs
Ephemeroptera	Mayflies
Odonata	Dragonflies
Psocoptera	Book-lice
Anoplura	Bird-lice and Sucking Lice
Thysanoptera	Thrips
Hemiptera	True bugs—Stink Bugs, Water Boatmen
Homoptera	Scale insects, Aphids, Leafhoppers
Neuroptera	Alderflies, Snakeflies and Lacewings
Trichoptera	Caddisflies
Lepidoptera	Butterflies and Moths
Coleoptera	Beetles and Weevils
Hymenoptera	Ants, Bees, Wasps and Ichneumon Flies
Diptera	True Flies
Aphaniptera	Fleas
Isoptera	Termites

Collecting Insects

Insects may be caught on the wing and may be looked for on flowers, under bark on trees, in rotten wood, in decaying animal and vegetable matter, as well as under stones and fallen leaves, especially when moist. Beating and sweeping bushes can dislodge many insects. Ground beetles may often be caught by making a trap using a small glass or plastic jar buried up to the rim in the ground and baited with meat. Moths, some beetles and other insects can be collected by luring them to a bait after dark. Brown sugar mixed with beer or rum makes good bait, and an apple cut open and smeared with sugar is effective. A light trap consisting of a lamp against a white background is useful for catching moths. Hanging carcasses of animals often attract insects, including butterflies which are also attracted to strong cheese. Fleas, lice and other parasites are obtained from the bodies of freshly killed animals. Dead refrigerated animals do not harbor infestations because of their low temperature. Birds' nests and the burrows of moles and rabbits are often fruitful places for collecting insects.

After the insects are captured and killed, they may be kept in a Riker mount. This may be any small flat tin. At the bottom of the tin, place a few crystals of para-dichlorobenzene or naphthalene, and cover the crystals with some cotton taken straight from the roll so that it is quite flat. The dead insects are carefully arranged on this cotton, and the lid is pressed shut. Glass-topped boxes may be used, and the insects shown together with their larval stages and food plants, in this way.

Insects are considered to be the most successful of all animals. They have spread to every corner of the world. That is to say, everywhere that will support them. This excludes, of course, the extreme polar regions. In forests we find insects living as predators, some eating the leaves of the trees, others sucking plant juices, gathering pollen and nectar, others scavenging, while some are parasites on other insects. Even here they do not compete with each other. Some plant-eaters choose grass, while others select leaves, wood, bark, or decayed wood.

Insects are limited in size by their system of respiration. Since air must travel by diffusion through the tubes, it is unable to reach the interior of large animals fast enough to support great activity. The other limiting factor is the chitinous exoskeleton, which makes flying difficult for larger insects.

Insects develop from eggs, most of them passing through a larval stage and then to a pupal stage before hatching as a perfect insect. This is called metamorphosis. Not all insects go through these stages; some simply hatch from the egg.

Most insects are readily identified by distinctive characteristics. The body is divided into three main parts—the head, thorax and abdomen. The head is usually in one piece, attached to which are the antennae. There are single and compound eyes, and variously modified mouth parts. The thorax is segmented and bears one or two pairs of wings and three pairs of legs. The abdomen is also segmented and, apart from the spiracles 'porthole like' openings to the tracheal system, is without appendages except for the posterior modified ovipositors (egg-laying organs).

Basics in collection and preparation are found on pages 15–17. Adult insects are usually pinned out in the dry state. There are special places where it is best to place the pin when mounting the insect. Beetles are usually pinned through the right wing cover or elytra. Flies and other similar insects including bees, wasps and ants are pinned through the thorax. The Hemiptera or bugs are pinned through the scutellum, a small triangle behind the neck. Some insects too small to pin may be stuck onto 'points', small triangular pieces of card which are attached by pins to the cork base of a box.

A cigar box fitted with a balsa wood base is a good way of keeping the collected specimens. The wings of butterflies and dragonflies may be laid onto a piece of card that has been rubbed over with a wax candle and then ironed with a hot iron. The shape of the wing and the scales will make a beautiful and interesting exhibit. The pressing can be varnished and kept with others in a filing system, using the back of the card to give the name and any other details.

Eggs and larvae of various insects, particularly butterflies and moths, can be found on their favorite food plants. The plant may be kept in a pot, or portions of the plant placed in a small glass container, to allow the stages to develop. In this way, the entire life history can be observed, and many of the various stages can be collected.

Giant
Water Beetle

Purple Emperor

Purple Emperor
Larva

Hornet

Giant
Dragonfly

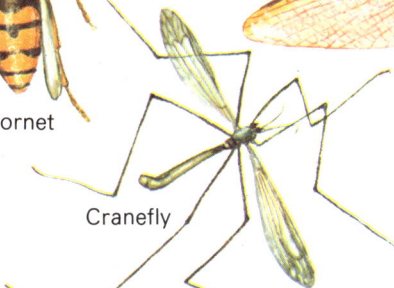
Cranefly

Ichneumon Fly

Rust-colored
Dancer Beetle

Great Caddisfly

Mud Bug

Earwig (Male)

Female Field Cricket

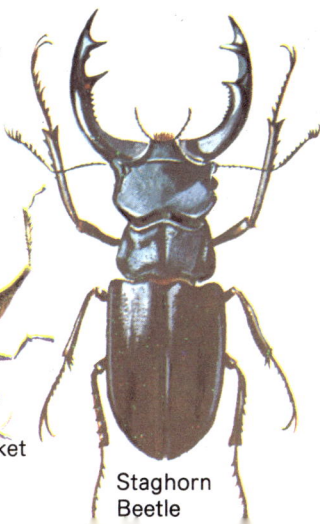
Staghorn
Beetle

Arthropods Other Than Insects

These animals include the crabs, shrimps, barnacles, woodlice, spiders, mites, scorpions, centipedes and sea spiders. All these animals have jointed and segmented bodies and most of them have a hard shell or external skeleton.

Crustaceans

This class of animals includes shrimps, prawns, water fleas, barnacles, crabs, lobster, sandhoppers, woodlice and crayfishes.

All are essentially aquatic and representatives of all these animals are found in the sea and in fresh water and damp places on land. Some are parasitic and unrecognizable as crustaceans.

Shrimp and water fleas may be collected, briefly immersed in 70 percent alcohol and then dried out. Barnacles need to be relaxed in dilute alcohol (30 percent) before preserving in 5 percent formaldehyde or 70 percent alcohol. Crabs and lobsters are best cooked in boiling water to remove most of the flesh and the specimens are then dried out.

Woodlice, found under stones and fallen trees in woodland, may be collected and preserved directly in 70 percent alcohol.

Arachnids

This group includes the spiders, mites, ticks, scorpions, centipedes and horse-shoe crabs. They are mostly land dwellers except for some mites and horse-shoe crabs, and they are mainly nocturnal. Spiders are abundant in late summer and are easily discovered by finding their webs. They can be collected and preserved in 70 percent alcohol. Mites are found when pond-dipping and may be preserved in 5 percent formaldehyde. Small centipedes are collected from vegetation on the woodland floor and in cracks in the earth, and may be dried out after immersion in 70 percent alcohol. The Berlese extraction apparatus mentioned on pages 18 and 19 is used in this work.

Probably no group of animals is more disliked. There is good reason for this as some of the spiders are poisonous, while mites are parasitic on human skin, suck blood and spread disease. However, they are an interesting group of creatures to study.

Crustaceans

Blue
Crab

Millipede

Centipede

Spiders

Echinoderms (Echinodermata)

Starfish and sea urchins are the most familiar echinoderms. The radial symmetry of these creatures and the firm outer skeleton are typical features. However, some, such as the sea cucumbers, resemble large slugs and others look like branching plants. Echinoderms are found on beaches below high-tide marks and may lie exposed in pools or buried in sand. Many seek the shelter of rocks, stones and seaweeds, and some burrow into sand and mud. Some echinoderms live at great depths in the sea.

Except for the sea cucumbers, echinoderms can be dried after previously preserving them in 70 percent alcohol. Sea cucumbers are best preserved, after relaxing in fresh water for a few hours, in 70 percent alcohol.

Starfish

Starfish common to the North American coast are of many interesting varieties. The Common Starfish normally has five arms, although some have four or six. They vary in color from yellow, orange and brown to a pale violet. The Spiny Starfish is smaller than the common variety and has knobs and spines on its arms. It is a pale-colored animal with a yellowish or greenish tinge. Cushion stars are small, flat starfish, yellow-green in color and about an inch across with short arms. Brittle stars have a body resembling a disc usually no more than a half-inch to an inch in diameter. Their long,

Spiny Starfish   Feather Star

thin arms emerge from the top of the disc and are usually five in number. These green or grayish-brown animals move by jerking two limbs forward at a time.

Sea Urchins

Sea urchins are round, globular-shaped animals which look much like pin cushions. They are covered with movable outward-pointing spines which may be poisonous. Their brittle skeleton is covered with projections to which the spines are attached and rows of holes through which tube feet protrude. Sea urchins live along rocky shores, feeding on algae and decaying materials. They are important as one of the scavengers which keep the ocean floor from becoming fouled. They provide food for many marine animals and are especially favored by gulls which will pick them up, fly aloft and drop them onto the rocks in order to break their 'shells.' They may also be used as food for man, and some produce dyes.

Sand Dollars

Most people are familiar with the pancake-like skeleton of this animal. Like the sea urchin, a close relative, sand dollars in the living state are covered with spines and have small tube feet which aid in locomotion. These animals swallow sand and digest the organic material in it. They are

Common Urchin   Long Spined Urchin

common on the Atlantic coast, north of New Jersey and on the west coast, south of Puget Sound.

Feather Stars and Sea Lilies
These animals belong to a group of echinoderms known as the crinoids. Sea lilies appear as flowers on a limey stalk. Being deep-water forms, they are seldom seen. Feather stars inhabit shallower waters and hence are a bit more familiar. They swim about by waving their feathery arms. When young, they are stalked also. Fossil evidence seems to indicate that these creatures were a great deal more common during Paleozoic times.

Sea Cucumbers
Sea cucumbers are fleshy, cucumber-shaped animals with a thick, leathery skin. Most of the time they creep about on the ocean floor by muscular movements of the body wall; their tube feet are of little use. They ingest sand and mud and digest the organic matter out of it. In some areas, particularly the Orient, they are dried, smoked and used in making soup.

Sea Cucumbers

Hemichordata
This group includes acorn worms and forms resembling jelly-fish, the pterobranchs. Sea dwellers, these occupy the same

habitat as the sea worms and other similar creatures. They are often dug up in the same places as sea worms. They are fragile and must be preserved on collection in 70 percent alcohol, relaxing first if necessary in seawater containing a few grains of menthol.

Balanoglossus, the Acorn Worm, and other related animals are widely distributed in shallow seas, where they live in the mud in U-shaped burrows. They strengthen the walls of the burrow, using mucus secreted from the skin. When feeding, sand and mud is swallowed and its organic content digested. The wastes form a cast which is ejected from the burrow.

Balanoglossus is a worm-like animal of the Hemichordata

Chordates (Chordata)

This group includes the sea squirts, salps, lancelets and the vertebrates, or animals with backbones.

Sea squirts are found in shallow water and quite often in rock pools on the shore, where they may be found attached to rocks or crevices. Water is drawn in through the mouth and sieved through gill slits to extract the organic material for its food. These marine animals are relaxed in fresh water and preserved in 5 percent formaldehyde.

The Lancelet, Amphioxus, has very wide distribution in shallow seas, but it lives only in the purest sand or gravel. It is so named, because of its lancet-like (surgeon's knife) shape. It has a rapid fish-like movement in the water and can travel just as fast through gravel or sand, provided that it is covered with water. The food is taken in with water and strained in a similar fashion to the sea squirts. Breeding takes place in early summer and is so precise that the shedding of the sperms and ova takes place within 24 hours. Lancelets are preserved in 70 percent alcohol.

Amphioxus

Sea Squirts

The Vertebrates

The majority of animals with which we are most familiar are vertebrates. Vertebrates are bilaterally symmetrical animals with a spinal column of vertebrae which protects the dorsal nerve cord and an internal skeleton. They have a head with a brain encased in a skull. Most have four limbs.

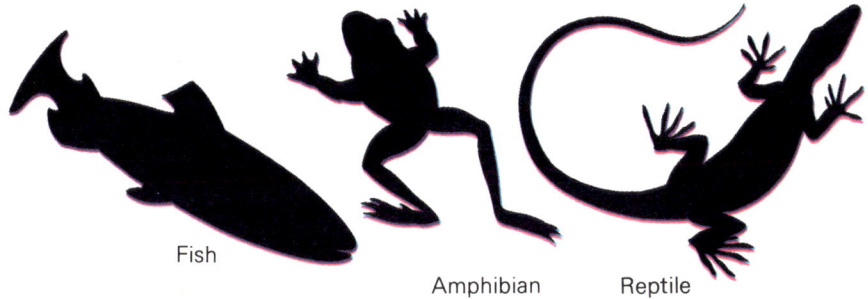

Fish   Amphibian Reptile

Fish may be fresh-water or marine. They are classified in two main groups—cartilaginous fishes and bony fishes. Cartilaginous fishes have skeletons composed of cartilage and include sharks, skates and rays. Bony fishes have bony skeletons. There are many families of bony fishes. Fishes breathe through gills for their entire life.

Amphibians only utilize gills for a part of their life and metamorphosize into lung breathers for the remainder of it. They may be fresh-water or terrestrial in habit although most return to water for reproductive purposes. Their skin is quite thin and usually slime-covered—if it dries out, they die. They have no claws or other protective structures and must flee or hide from danger.

Reptiles are terrestrial or semi-aquatic. They utilize lungs throughout their lives. Their bodies are covered with scales, and those that have feet have claws on their toes. Their eggs are not gelatinous, as in the two groups just discussed, but are tough and leathery.

Birds also are lung breathers. They, too, have clawed feet, but their scales are modified into feathers over a large part of their bodies. In most species, the forelimbs are modified into wings which are used for flight. Bones are hollow and air-filled, and eggs are encased in a shell. Most are incubated by

77

the parents. These are the first of the warm-blooded animals.

Mammals have the most highly developed nervous system. They are warm-blooded lung breathers with a hairy body covering. Young are born alive and nursed by the mother after birth. Her milk-producing mammary glands give the group its name.

Bird

Mammal

Fishes

Cartilaginous Fishes

Lamprey: This fish is snake-like in appearance and has no jaws. The mouth is round with horny teeth and a rasping tongue. Lampreys attach themselves to fish by their sucker mouths and rasp away the flesh with their tongues. Other species of lampreys include a smaller river lamprey and a still smaller brook lamprey.

Sharks: Sharks range the oceans of the world. Many supply oil, livers used medicinally and a waterproofed, transparent tissue. Some are dangerous because of the blow of their tail and their bite. Dogfish are small sharks, usually about two feet in length. They have a skeleton like the lampreys, but also have some spines in their skin called dermal denticles. Mackerel sharks are surface feeders which grow as long as 12 feet.

Skates and Rays: This group includes skates, sting rays and the torpedo, or electric, rays. Torpedo rays have rounded bodies. Their electric organs produce a shock strong enough to knock a man over. Sting rays have the typical flattened body of a bottom dweller plus a poisonous spine on their whip-like tail. Similar in appearance, but differing in that they do not bear live young, are the skates. Their leathery, H-shaped egg cases are often washed ashore.

Lamprey

Greater Spotted Dogfish

Porbeagle (Mackerel Shark)

Thorny Skate

Bony Fishes

Herring: Most important of the fishes, economically, is the herring family, which includes, along with herring, the shad, sardines, pilchard, alewife and menhaden. The first four are important as food for man while the alewife is an important food for other fishes such as trout. The menhaden catch is probably the largest of all, over a billion fishes a year, but few are used directly for food—most are converted to fish meal, oil or fertilizer. Making it easier to catch these fishes commercially is the fact that they congregate in schools and can be netted in quantity.

Common Carp

Minnow

Cod: The cod is one of our most important sea fishes, being second only to the herring in importance. It is a large fish, and can reach 5 feet in length and weigh 40 pounds. Other members of the family include the haddock, coalfish, whiting, pollack, ling, hake and the rocklings.

Trout: Trout range from lake trout which live in the depths of great lakes and have been caught weighing as much as 102 pounds to some which, salmon-like, live in the sea and return to fresh-water streams to spawn. Some species which may lead salt or fresh-water lives include the Rainbow, Cutthroat, Brown and Brook. Many states maintain extensive hatchery programs for stocking waters with members of this very important fish family.

Minnows: This family includes the carp, goldfish, cat-fishes, dace, chub, shiners and minnows, 193 North American species in all. Most of these fresh-water fishes have little value other than as bait and food supply for larger fishes. Goldfish were domesticated by the Chinese a thousand years ago. As bowl fishes they are attractive, but in the wild they revert to dull, carp-like coloring. They should never be released as they are a threat to native fishes and quickly become pests. The whiskered catfish is probably the most tenacious of fresh-water fishes. They live longest in warm or polluted water.

Eel

Sunfishes: Of all fresh-water fishes, the sunfishes are probably best-known, at least east of the Rockies. It includes Bluegills, Pumpkinseeds, Crappies, Rock Bass, Warmouth, Small-mouth, Largemouth and Spotted Bass. The members of this family range from a 3-inch sunfish to a 22-pound, large-mouth bass.

Eel: This snake-like fish is a common inhabitant of fresh water. Fresh-water eels have an amazing life history. The young swim great distances from their spawning grounds in the Sargasso Sea in the Atlantic Ocean, to reach fresh water in land bordering the Atlantic Ocean. They remain in fresh water for several years, but return eventually to their place of hatching where they spawn and die.

Flatfishes: This unique group is common in both Atlantic and Pacific waters. They have developed a very unusual body form adapted to life on the ocean floor. They actually swim on their side but both eyes are on the 'upper' one. When young, they have one eye on each side, but as they age one eye migrates to the other side. In addition, the bottom side is whitish while the top side varies in color. A flatfish is able to change its colors and pattern to match its surroundings. Several of these fishes are valuable as food: flounders, halibuts, turbots and soles.

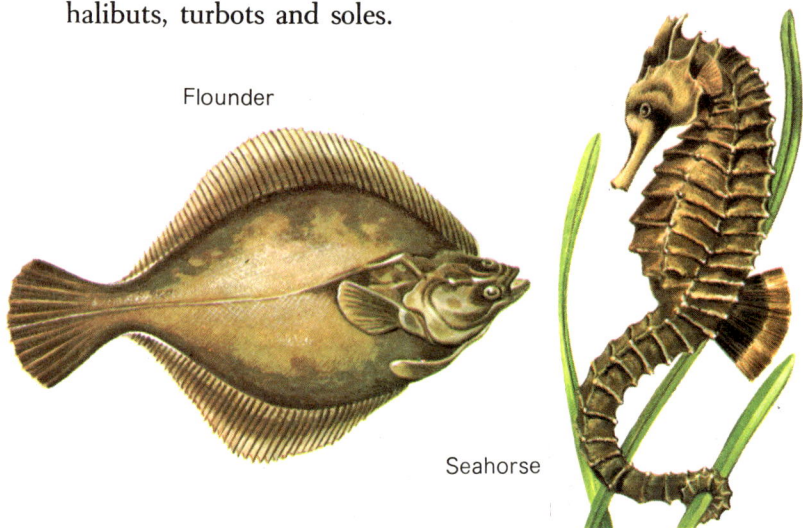

Flounder

Seahorse

Seahorses: Many people do not think of these strange creatures as being fish due to their un-fishlike appearance. Their bodies are covered by thin bony plates. Seahorses swim upright and cling to vegetation with their prehensile tail when they wish to rest. Breeding habits are very interesting—the females deposit eggs in a pouch on the underside of the male where they remain until they hatch. Most seahorses live in shallow southerly waters.

Since there are 4,000 species of fishes listed for North America, this has only skimmed the surface of the more common families. Many of the deep-sea food fishes such as Tuna, Bonito and Swordfish have not been mentioned nor have the common aquarium fishes such as Guppies, Swordtails, Neons and Angelfish. Space does not permit elaboration.

Aquariums

The fishes described above are mostly sea fishes and are best observed in their natural habitat, where possible, as it is difficult to keep sea fishes in an aquarium. This is not true for freshwater fishes, and those found in temperate regions can be easily kept in a small aquarium. Small fishes, such as minnows and sticklebacks, can be caught with a net in ponds or streams and transferred to an aquarium. Sticklebacks are interesting fish to keep in an aquarium, as they build nests.

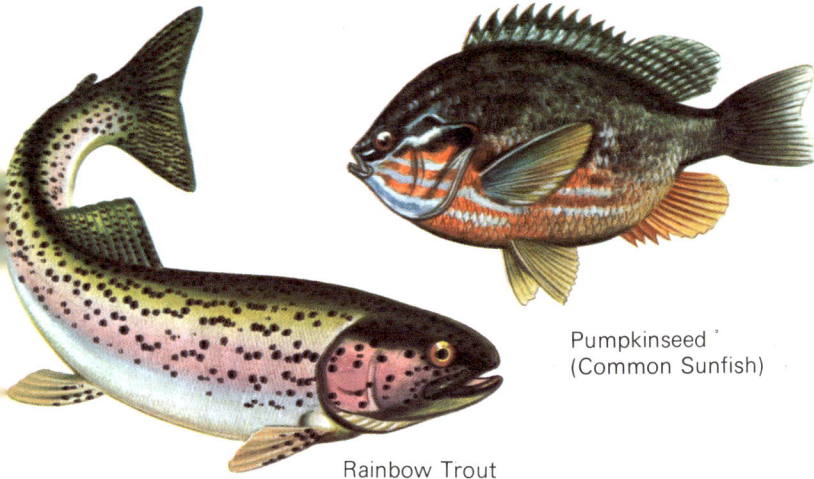

Pumpkinseed
(Common Sunfish)

Rainbow Trout

There are two main types of aquariums, fresh-water and salt-water, or marine. These may be classified as large or small aquariums according to the size of specimens kept therein, pure or mixed culture aquariums, and cold-water or warm-water aquariums.

All-glass aquariums are cheaper than those constructed with angle-iron frames, but they are broken more easily. A cracked pane on one side of an angle-iron aquarium can be replaced fairly easily, but a cracked glass tank is not only useless but dangerous if large quantities of water are being handled.

Animal and Plant Requirements
It is impossible to lay down hard and fast rules as to the number of fishes or other animals to be kept in a given volume of water. Normally a considerable amount of water is needed because most inhabitants of an aquarium require dissolved oxygen in the water for respiration, the exception being some pond dwellers. Hot and stagnant water contains a small amount of dissolved oxygen, whereas running cold water has a rich supply of oxygen. Trout, for example, need cold, running, well-aerated water. Siamese Fighting Fish, on the other hand, need only a small jar to live in, as they possess an organ that enables them to gulp air from the surface. There should be a gallon of water for every inch of fish.

In order to obtain the conditions of natural surroundings, it is necessary to 'balance' the aquarium; this is especially necessary with fresh-water aquariums, which are balanced by adding water plants such as the following:

AERATING PLANTS

Cold water	*Sagittaria*	Sword plant
	Vallisneria	Tape grass
	Elodea	Canadian pond weed
	Myriophylum	Water milfoil
	Fontinalis	Willow moss
Warm water	*Cryptocoryne*	Water trumpet
	Marsilea	Water four-leaved Clover
	Elodea	some varieties
	Cabomba	Fanwort
	Vallisneria	some varieties
	Echinodorus	Amazon sword plant

PLANTS THAT FLOAT ON THE SURFACE

Cold water	*Azolla*	Water fern
	Lemna	Duck weed
	Riccia	Crystal wort
	Hydrocharis	Frog bit
Warm water	*Salvinia*	Water fern
	Pistia	Water lettuce
	Eichhornia	Water hyacinth

Fresh-water aquariums are able to provide endless enjoyment and knowledge of fishes. If lights are placed behind the tank, the aquarium becomes even more attractive.

Illumination is sometimes critical in an aquarium, and in certain conditions it is essential to have plants which will flourish under controlled conditions of light. It is best to plant the plants in small pots of sand and earth and then arrange them in the bottom of the tank before any animals are added. The tender shoots are often eaten by the fishes, and putting plants in ahead of time will allow for some growth to take place before the fishes are added. All plants, as well as other materials used should be carefully washed to avoid contaminating the tank.

There are no weeds that can be used successfully for more than a short time in a marine aquarium. A few small red and green algae can be used but must be replaced every few days. Aeration is essential in marine aquariums. As simple a method as a fine stream of water splashing on the surface may be used with fresh-water aquariums, but one of the commercial aerators will be necessary for this use.

Several kinds of marine aquariums can be set up. One might present a collection of seashore life, including anemones, periwinkles, mussels, sea urchins, starfish (be careful, they eat mussels!), seahorses, small lobsters or shrimp. Another collection might contain animals of the deeper water—feather stars, fan worms and sea squirts.

Many marine animals can be fed directly, but some must be provided with plankton as they feed by filtering the water. For these, it is best to add freshly obtained sea water from time to time. The salt level of the aquarium must also remain fairly constant. If fresh salt water is not available, artificial sea water may be prepared using sea salt in a 4% solution. This can be gotten at pet shops.

There are two main problems encountered with aquariums. *Cloudy waters* occurring in fresh-water aquariums are a sign of the presence of bacteria, due in most cases to dead animals. Make sure first that this is not debris stirred up by the fishes. Cloudiness is often normal in a marine tank. *Green water* denotes the presence of algae. This does no harm, but makes any observation difficult; the remedy is to cut down the illumination. Do not overdo this or brown algae will start to form. Brown algae will occur whenever there is not enough light for a marine tank and proper balance can only be obtained by patience and experience.

Instead of an aquarium, there are many types of pools which can be purchased or constructed in one's garden. A very satisfactory one can be made from a plastic sheet.

After a suitable hollow is dug, a plastic sheet is placed in it. The edges are weighted down with rocks or soil and the water added.

Amphibians
Frogs and Toads

Bullfrogs, Spring Peepers and the American Toad are representatives of the orders of amphibians called the Anura or Salientia (leaping animals). They have no tail in the adult stages and are exclusively fresh-water and land animals. None can live in the sea, and this fact explains their absence from islands in oceanic areas of the world. They are distantly related to the lungfishes which can survive for months out of water. All these amphibians pass through an aquatic larval stage as tadpoles, and for a time they are entirely dependent on the use of gills, in the same way as are the fishes.

American Toad: The warty-skinned American Toad or one of its close relatives can be found from Canada to Mexico. It spends most of its life on land, preferring moist woodlands, gardens or lawns. Toads do not leap like frogs, but tend to move in short little hops. They often rise up on their stubby front legs to stalk their prey (worm or insect) until close enough for a thrust of their sticky tongue to capture it.

The American Toad usually breeds in a ditch or pond which is the home of generations of toads. The animals pair and clasp each other in the water, the male riding on the female's back. The eggs are laid in long strings and fertilized immediately. These strings, unlike those of the frog, remain separate and are often laid over stones and reeds; some of them never reach the water, drying up and dying. The mortality of these eggs is amazing, but so too is the invasion in early summer of tiny hatched toads, as they leave the water and move in thousands along rutted tracks and ditches, feeding on mosquitoes and small flies, snapping at them as they pass.

It is three years before the toads reach maturity; to do so they have to triumph over many problems. There are two things that come to their aid—they can stay motionless for weeks in a damp hole feeding on occasional small insects or worms which may drop in, and they have a warty skin in which there are poisonous mucous glands.

North America has 36 species of frogs divided into two families: the treefrogs and the true frogs. Treefrogs are generally small, slim-waisted and long-legged. As their name implies, they are adapted to climbing trees. Adhesive discs on the end of their toes enable them to climb or cling to vegetation. Many of them also change color. This ability seems to be affected by light, temperature, moisture and the frog's activity as well as its surroundings.

Spring Peeper: Seldom seen, but frequently heard, the Spring Peeper, one of the smallest treefrogs, seldom is over an inch in length, yet it has a voice which seems to come from a creature many times its size. It is easily identified by the X-marking on its back which explains a scientific name of Hyla crucifer.

Leopard Frog: Because of their wide distribution and common use in biology laboratories, the Leopard or Meadow Frog is probably the best known of the frogs. Its bright coloration and spotted appearance provides excellent cam-

Common Toad

ouflage both along the water's edge and in open grassy areas where it goes in search of insect food.

Bullfrog: The Bullfrog is the bass of the frog family. Its resounding call can be heard some distance from the large pond or lake in which it makes its home. Most aquatic of those mentioned, bullfrogs seldom range beyond the pond edge. Formerly found only east of the Rockies, the Bullfrog has been introduced to parts of the west coast and seems to be thriving. These large frogs, growing to eight inches in length, have proportionately large appetites. They will eat almost anything their size or smaller which ventures close enough: frogs, fish, birds or small animals as well as insects. In turn, they are sought by herons, snapping turtles and man. These provide the frogs' legs of commerce.

Frog development, from spawn through the tadpole stage to the adult.

Salamanders and Newts

Mud Puppies, Tiger Salamanders, Spotted Salamanders and Red Efts are some members of the order Urodela or Caudata. They are permanently tailed amphibians, retaining their tails after the larval stage has ended in contrast to the Anurans wherein the tail is slowly absorbed. The Urodeles spend much less time in water than most of the Anurans do and some remain on land all the time, with the exception of the breeding season. Their eggs are often laid separately, and a great deal of care is taken to select suitable leaves on which the egg packets are placed. There is no parental care of eggs either by Anurans or Urodeles.

North America has more kinds of salamanders than the rest of the world combined. Of eight known families in the world, seven are native to the United States—about 135 kinds in all.

Mud Puppy: The Mud Puppy may measure as much as 18 inches in length. These rather grotesque aquatic creatures are sometimes caught by startled fishermen. The Mud Puppy has three tufts of fluffy red gills behind a head flattened at right angles to a fin-like tail. It is dark in color, mottled and slimy. In spite of its size and unpleasant appearance, it is quite harmless. A Japanese relative grows up to five feet in length.

Tiger Salamanders: These salamanders belong to the mole salamander family. The family name derives from the fact that members live largely underground except during the breeding season. These dark creatures, marked with olive or yellow bands, are hardy and, in spite of their burrowing habits, make good terrarium pets. In some parts of Mexico and the western United States, this animal may retain gills in the adult state. This form is known to biologists as the Axolotl.

Spotted Salamanders: These are slightly smaller than a Tiger Salamander, reaching a maximum length of eight inches. They do not burrow as much as their relatives and may be frequently found in woods or meadows near marshes. They feed on insects, snails, slugs and worms. These hardy creatures have been known to live as long as 24 years.

Red Eft: This is actually the immature, land form of the newt. It is usually found among moist plant growth where it spends from 1 to 3 years. While on land, this creature is a bright red-orange spotted with black, but by the time it returns to an aquatic habitat to spend its adult years,

Newts always return to the water to breed, spending the rest of the year in damp places on land.

it has become a yellowish-brown, marked with red and black spots. These creatures destroy great numbers of mosquito larvae and pupae.

Another salamander family, the lungless family, includes the Northern Dusky, Red-backed and Slimy Salamanders, all of which are unique in that they open their mouths by lifting the upper jaw instead of dropping the lower one and they depend solely on skin respiration. The brook salamander family includes the Northern Two-lined and Long-tailed Salamanders which frequent streams and small bodies of water. Most uncommon are the blind salamanders which have lived so long in darkness that they have lost their sight and become almost transluscent. They are found only in cave or well depths and little is known about their lives.

Things To Do With Amphibians
Most amphibians can easily be kept in terrariums and aquariums. It is often possible to collect their eggs in the spring. These should be kept in a wide-mouthed container, out of direct sunlight. Toad and frog tadpoles are vegetarians and may be fed small bits of lettuce, cornmeal or cooked egg yolk. As they mature, provisions must be made for them to get out of the water. In most cases they can be transferred to a terrarium. Remember that frogs have a thinner skin than toads, therefore great care must be taken to pro-

Mud Puppy

Tiger Salamander

vide them with sufficient water to keep them from drying out. Frogs should have a water area in which they can swim, while shallow pans or saucers of water are all that are necessary for a toad.

Common toads make interesting and charming pets. Given a broken flower pot to live in and a free run of the garden, they will do a lot of good by eating insect pests. They do, however, have a tendency to stray, and perhaps it is better to keep them in a pen or area from which they cannot escape. Toads can burrow in soft earth. These animals easily remember people and will come to be fed. They appear to have a high standard of intelligence.

Toads are nocturnal and valuable friends to gardeners. They destroy slugs, worms and other insects. Toads will travel many miles each year to return to the pond or ditch where they were hatched. They will overcome obstacles in their path with a tenacity which is very impressive. They are often found walled up in rock crevices, and this has given rise to the untrue suggestion that toads can survive for hundreds of years in these conditions. Small insects will often crawl into a crevice and furnish the animal with food, and water globules from condensation will give the damp conditions necessary for life.

Newts can easily be kept, and the breeding habits of these interesting animals are most instructive. It must be remembered that the newts normally leave the water after breeding,

Newt

Young or Red Eft

Dusky Salamander

Adult

and they must not be left in deep water after this time. They can be fed on small worms or tubifex worms used to feed aquarium fishes. They will also eat tadpoles. It is important to remember that their skin must be kept moist. They can be killed by being held too long in a warm, dry hand.

Newts are most instructive creatures to collect and watch. The larger example of this group, the Axolotl, or Mexican Salamander, is a great favorite among aquarists. This curious creature may be 6 inches in length and can spend its entire life in the larval stage wherein it has large external gills. Sometimes, as a result of heat and drying of a pool, or an excessively hot classroom where the creature is kept, the Axolotl will lose its external gills and develop eyelids and lungs so that it can live entirely on land like other amphibians, such as frogs and toads.

Vivariums are artificial enclosures for raising or keeping animals, especially such creatures as terrapins, newts and other semi-aquatics.

Reptiles
The reptiles of North America are classified into three main groups: lizards and snakes, alligators and crocodiles, and turtles and tortoises.

Modern reptiles are cold-blooded vertebrate animals which have neither hair nor feathers. They always breathe by the use of lungs and do not undergo a transformation called metamorphosis, from gill breathing to lung breathing, as do amphibians. The skull of a reptile is attached to the vertebra of the spinal column by a single knob or condyle and this feature is shared with the birds. The amphibians have two condyles, as have the mammals.

When we say the reptiles are cold-blooded, it does not mean that their temperature is always low. Their temperature rises and falls according to their surroundings. This is why a lizard is so active in the hot sun on a log, and yet will not move at all on a cloudy, windy day.

Reptiles can be said to have once ruled the world. They were so big and powerful with seemingly limitless possibilities to spread over the world. However, at the present time they are in the wane of development. All animals have had their heyday and the age of reptiles seems to have been one of the most successful.

Lizards and Snakes
Most lizards look much like salamanders with a scaly armor. They are quite different in habit, however. As much as salamanders seek dark, damp areas, lizards thrive in warm, dry, sunny areas. The majority of North American lizards are to be found in the arid southwest.

Horned Lizard

Fence Lizard

Horned Lizards: Many children are familiar with these lizards, often sold in pet shops. There are 13 species in the United States, most of which live in areas which seem too dry and sandy for life to exist. These round, flattened creatures, covered with spiny projections, are quite harmless and have some unique habits and abilities. Perhaps strangest of all is their ability to shoot blood from their eyes when frightened. If kept as pets, they must be kept in a warm, sunny spot or they will not eat.

Swifts: The swifts include the Scaly, Spiny and Fence Lizards. These are the most widely distributed, ranging from coast to coast, Canada to Argentina. Swift, agile lizards, they are characterized by shingle-like, overlapping scales which have a keel-like spine. The largest is about 5 inches long. Insect eaters, they are good climbers and well protected on fence post or tree trunk by their coloration and skin texture.

Gila Monster: Fortunately, the only poisonous lizard of the United States has a range limited to the desert areas of the southwest. It is quite large, up to two feet in length, with a fat, stubby tail. Imbedded in its skin, tiny round bones give it a beaded appearance. Its venom is only used for defense, its teeth are adequate for food-getting—favorites are birds eggs, lizards and small rabbits.

Glass Snake: This creature is not a snake at all, but a limbless lizard. Like other lizards, it has eyelids, ears and a scaly belly. Also like many of its relatives, it has the ability to shed its tail. The renewed tail, a sac of flesh, lacks bones and often seems to have been stuck on the animal. Glass

Garter Snake

Snakes, or Joint Snakes, as they are sometimes called, are found in the central and southern United States and parts of Mexico.

Snakes are the most widely distributed as well as most numerous of the reptiles. There are 126 species in the United States but only 22 in Canada. Being cold-blooded creatures, it is easy to understand why there are more in tropical regions. Many scientists believe that snakes evolved from burrowing lizards. Certainly there does seem to be a possible transition from the glass snake to the true snakes.

Common Garter Snake: This is one of the most widely distributed and frequently encountered of the North American snakes. It lives in a variety of habitats: fields, bogs, marshes, gardens, stream and pond edges. Its food is primarily earthworms, salamanders, frogs and toads. It is one of the earliest snakes out of hibernation in the spring and latest to retire in the fall. Its color and markings vary greatly, but there is usually a longitudinal striping effect. These snakes are harmless and make excellent terrarium pets.

Racers and Whipsnakes: Members of this snake family are found throughout the United States. They have slender bodies and are fast moving. These snakes eat their prey — small mammals, birds, frogs, lizards, snakes and insects — live; they neither poison nor constrict it. The Black Racer of the eastern United States attains a length of 4 feet and may climb trees to escape danger. Whipsnakes depend solely on their speed to escape.

Scarlet Kingsnake

Black Racer

King Snakes and Milk Snakes: This family has representatives in most of the states and southern Canada. These 3- to 4-foot long constrictors frequent barns and stables seeking rats and mice, not milk. Members of this group are known as rattlesnake killers. Unfortunately, the Milk Snake is marked very much like the Copperhead is and many of these harmless, actually beneficial, snakes are killed needlessly.

Copperhead: A large proportion of bites by venomous snakes in the eastern United States are inflicted by the copperhead. Fortunately, it is nocturnal in habit or the incidence of bites might be higher as it exists in populated areas. Copperheads are pit-vipers as are water moccasins and rattlesnakes. All of these have heavy, triangular heads, pits which are sensitive to temperature gradations, hypodermic-like fangs which inject poisons upon biting, and elliptical, cat-like eye pupils. If you can observe these characteristics, you are probably too close! A Copperhead may be identified by reddish-brown triangular patches on its side having the base of the triangle by the ground. Flattening its skin would give the appearance of hourglasses along its back.

Rattlesnakes: There are about 15 species of rattlesnakes distributed throughout the United States. They are a bit easier to identify and avoid due to the rattles, a series of loose, dry segments at the end of their tails. These segments are added to with each shed and do not indicate the age of the snake. Noise produced by vibration of the rattles is a whirring sound. Many rattlesnakes are kept caged and 'milked' of their venom for the purpose of producing antivenins.

Crocodilians
Only two members of this group occur in the United States:

Crocodiles: These are more aquatic than alligators are. They are grayish-green with a triangular head and pointed snout having the fourth tooth of the lower jaw fitting into notches in the upper jaw. They are restricted to the marshes of southern Florida.

Alligator: The American Alligator is found in rivers or swamps from the Carolinas southward and as far west as

Copperhead

Rattlesnake

the Rio Grande. At one time, these creatures were plentiful, and 20-foot long specimens could be found. Excessive harvesting for hides and widespread capturing of hatchlings has endangered this species. Legislation now prohibits the taking of 'gators in most of their range, but poaching continues to be a problem.

Alligators spend much time basking in the sun or lying half-submerged in the water. Projecting eyes and nostrils on up-turned snouts enable them to float almost completely underwater while still breathing freely and looking about. In this position, their scaly backs look like a log's rough bark and all they need do is wait patiently for an aquatic animal to approach, at which time, the 'log' comes to life and an unwary creature becomes a meal.

Turtles and Tortoises

Turtles are probably more like their ancient ancestors than any of today's animals. They seem to have changed little in millions of years. These armored reptiles live longer than any other animals with backbones. Some tortoises are known to have lived 150 years. The term tortoise is used for land-dwelling turtles while the term terrapin, Indian in origin, is used to refer to those turtles more commonly used for food. Most tortoises, vegetarians, have strong feet adapted for walking and claws for digging and high, domed shells. Water turtles, on the other hand, eat a variety of

Alligator

food, their feet are webbed to flipper-like, in ocean species, and their shells are more flattened. It is the shell, a case into which vulnerable body parts may be withdrawn, which has enabled the turtle to survive for millions of years.

Box Turtles: These are probably the best armored of the turtles, having both front and rear of their plastron, the bottom of the shell, hinged. This permits them to form a tightly sealed 'box' after head and legs are withdrawn. Because of the high-domed shape, the Box Turtle is not a very good swimmer and may drown if overturned in the water. It is a land dweller and eats a variety of food.

Snapping Turtles: Snappers, along with close relatives, the mud and musk turtles, are only found in the Americas. The Common Snapper is the most widely distributed. Very seldom does it leave the water. Its terrible disposition probably is due to the fact that its limited armor makes it more vulnerable than other turtles. Common Snappers of 20–30 pounds are dwarfed by the Alligator Snapping Turtle which has been found as large as 219 pounds.

Painted Turtles: Although they are not found in parts of the southwest or southern Florida, the Painted Turtle is probably the most common turtle in the United States. It is one of the first to emerge in the spring. Basking in the sun on log or rock, it will slide into the water at the slightest hint of danger.

Snapping Turtle

Painted Turtle

Things To Do With Reptiles

Of all the reptiles, turtles are probably most widely kept as pets. Unfortunately, many of the hatchlings sold at pet shops are in such poor condition that they soon die. Often these turtles are southern varieties: Mississippi Map Turtles, Red-eared Sliders or Southern Painted Turtles. These need much warmth and light. They should be kept at a temperature of 75° to 85° and have 8 hours of bright light but little direct sunlight. Turtles eat a variety of food from raw meat to fruits and vegetables. Dried 'turtle food' alone is not adequate. Larger turtles need more room than is available indoors. They may be kept in outdoor pens or pits.

Snakes and lizards also make good pets and may be kept in vivariums or snake cages. Great care must be taken in handling poisonous snakes. Snakes eat only animal matter and though many of them normally eat live food, they can be taught to eat dead food. If a snake refuses to eat for several weeks, it should be released.

Due to their endangered status, alligators can no longer be obtained as pets. Several of their relatives are available, but because of their nasty dispositions, they do not make good pets.

Reptile eggs may be easily hatched. They must be kept warm and moist. Reptiles are easily preserved in a formalin or alcohol solution. Snake skins can be cured by a simple salt and alum process or shed skins can be stretched and dried and then glued on pieces of cardboard. Photography of reptiles is quite challenging due to the speed and wariness of these creatures.

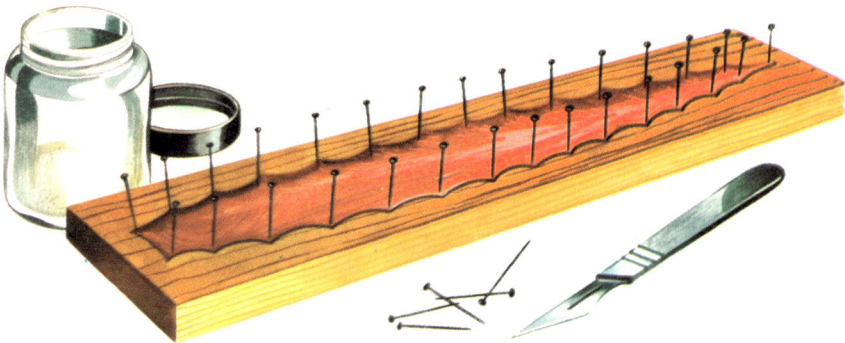

A stretching board is used to prepare small animal skins.

Birds

About 130 million years ago, in what is now part of Europe, there lived a strange bird-like creature. It was about the size of a pigeon, with teeth in its jaws and a long lizard-like tail.

It had feathers, at least in the tail region—this is known from fossil remains found in the Solenhofen limestone of Bavaria, Germany. The limestone, by the way, is used in lithography, where it is known as lithographic limestone. This fossil was called the 'Bird of the Dawn', or *Archaeopteryx*. This creature had legs covered with scales—birds of today have similar types of scales on their legs.

The chances of finding good fossils of any bird-like creatures are slim, as the conditions necessary for fossil formation are not suited to a flying animal.

A good fossil development requires still and quiet conditions and fine mud or silt in a shallow lagoon where a dead bird might lie undisturbed as the silt gently covers it, finally to harden into rocks recognizable today as limestone and other matrices. It would certainly account for the rarity of fossilized bird remains. This mainly applies to flying birds; flightless birds had a better chance of becoming fossilized in their entirety because of their ground-living habits. Many birds have gone the way of the *Archaeopteryx*. These include the Great Auk and the Passenger Pigeon. The incredible numbers of Passenger Pigeons made it almost inconceivable that the species could ever die out. In 1847 there was an account of a flock of them in a column 500

The Pheasant is one of the handsomest of birds

yards wide taking three hours to pass. It was estimated that as many as one billion birds were seen. The strange thing is that each bird laid only one egg but hatched several broods every year. These birds were found in the eastern United States where perhaps the cutting down of woodlands and indiscriminate shooting hastened their end.

There are approximately 8,600 different species of birds in the world, several hundred so rare as to be on the verge of extinction. There is a great deal of public sympathy toward the preservation of birds, and this is shown by the numerous societies and organizations dealing with the many aspects of birds and their conservation.

The well-known National Audubon Society was begun in an attempt to protect the egret. Other organizations on national, state, and local levels concerned with the protection of birds, as well as other aspects of our environment, include National Wildlife Federation, Nature Conservancy and the Sierra Club. These organizations hold meetings, provide programs and field trips in which you might participate.

Birds soon become used to human contact. There are very few of them that will not take an interest in whatever a homeowner is doing in his flower garden or yard. Some robins, for example, will perch on the garden fork when the spade is being used, and then transfer to the spade when the fork is being used, all in eager anticipation of the worm that will be exposed. The Starling will clean up after the gardener, removing cutworms, grubs and slugs with relish. Everyone

Owl

Falcon

Herring
Gull

will be aware of gulls who now seem to be making their homes in places far from the sea. It is this willingness on the part of some birds to exist in and near human habitation that may result in either their ultimate survival or their extinction.

Birds are classified in numerous ways, all based on specialized scientific information. They fall into two main groups, one of which is those which are incapable of flight — the ratite birds. These birds have no keel along the sternum and are ground living. The other group — the carinate birds — possesses a carina or keel along the sternum, allowing the attachment of the flight muscles and the general anatomy associated with flight. Examples of flightless birds are the ostrich, rhea and emu, found in Africa, South America and Australia, respectively. The flying birds have an enormous variety of shape, size, color and habit, varying from the giant soaring albatross and condor to the tiny hummingbird.

Things To Do With Birds
Federal government regulations clearly state that it is illegal to take or possess any migratory bird, its nest or eggs without special permits. Since there are very few exceptions to this law, it is best to limit birding activities to observation, photography and feather collecting. Egg collecting should never be undertaken, and birds should not be disturbed at the nest. Examination of pellets disgorged by owls and other birds of prey is a fruitful source of small mammal bones, skulls and also teeth. Kingfisher and heron pellets may yield fish bones.

Wren

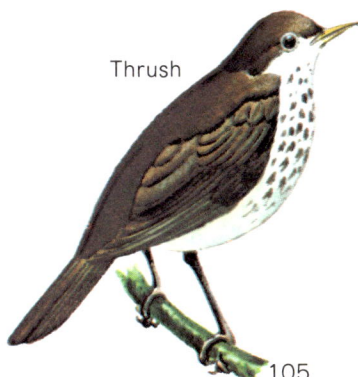

Thrush

Bird Watching

This is a perfect example of a useful collector's hobby, the collection and marshaling of facts about a most interesting local inhabitant of your garden or field. The need for careful notes is essential. Some form of attraction is often helpful in persuading the birds to come close enough for detailed observation. Bird houses and feeders in the garden, well-stocked window boxes or a few weeds left growing in one section of the garden will attract many interesting birds. In winter, brightly colored berries such as holly and the multiflora rose hips attract the birds. Also in the winter months, supplementary feeding will attract an abundance of birds.

Chickadees, nuthatches and titmice, being insect-eaters, are easily attracted by suet and peanut butter mixtures, but they also will eat sunflower seeds, nutmeats finely

Binoculars with a very wide field of vision are best, as they allow the bird's flight to be followed.

Blinds can be constructed of any material capable of providing camouflage.

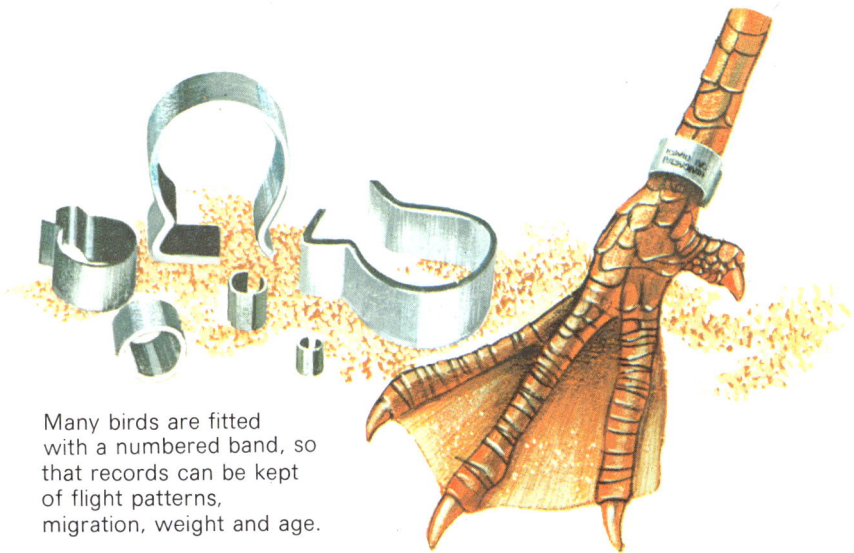

Many birds are fitted with a numbered band, so that records can be kept of flight patterns, migration, weight and age.

ground dog biscuits and stale doughnuts. Birds which eat fruits and berries, such as the Robin, thrush, Blue Jay and Mockingbird, may appreciate bits of oranges, apples or raisins. Seed-eating birds are the easiest to attract, but even they fall into two categories—some will eat from raised feeders (finches, jays and cardinals) while others (sparrows, juncos and Mourning Doves) are ground feeders. Most of these birds can be attracted by mixtures of sunflower seed, millets, grains, cracked corn, bread, cereals and popcorn. The best way to find out the food your visitors like is to put out many kinds and watch the results. Even individuals have varying tastes.

The National Audubon Society offers much information on both bird watching and feeding. Particulars can be obtained by writing to the Society's headquarters at 1130 Fifth Avenue, New York, New York.

Bird Banding

Since 1920, the banding of migratory birds in the United States and Canada has been under the joint direction of the federal governments of the two countries. In order to avoid confusion, all the bands used on birds in North America are issued by the United States Fish and Wildlife Service. A bander must keep accurate records when banding of the number on the band, species, age and sex of the bird, place and date of banding. Records are then sent to the Bird Banding Office.

Nuthatches, chickadees and titmice like artificial cavities.

When a banded bird is found, it should be reported to the same office. If the bird is dead, remove the band; return it with your name and address, information on the band, date and place found and the condition of the bird at the time. If the bird is alive, do not remove the band, but forward as much information as possible. The analyzation of these records provides much useful information on migration and other aspects of bird life.

The Annual Christmas Bird Census sponsored by the Audubon Society can be engaged in by window feeder watchers as well as experts. Hawk watches are carried on by many local clubs. Amateurs can be of great help during times of trouble such as the Torrey Canyon disaster, the Santa Barbara oil slick or unusual storm conditions. Even providing nesting boxes and feed is beneficial to birds.

Nest boxes

Many varieties of wild birds will lay their eggs in nest-boxes or any artificial sites in gardens, hedgerows or other places selected by the bird watcher. There are many different kinds of nest-boxes, many of which can be made at home.

Nesting ledges are preferred by the Robin.

Purple Martins are apartment dwellers

Feather Collections

Feathers collected can be mounted on pieces of cardboard or in a scrap book. Remember to keep careful notes of where and when they were found. Visits to zoos can provide unusual additions to your own collection, which may be arranged by species, type of feather or the location where the feather was found.

Preparing Bird Pellets

Pellets from birds of prey often contain bones. They may be collected from beneath trees in which the birds roost. Pellets should be boiled for a few minutes in water with a trace of baking soda. Any hair present will rise to the top, forming a mat, while the heavier bones and any debris present will fall to the bottom. The hair may be poured off and the bones given several rinses in running water. They may then be laid out on toweling or blotting paper to dry. Discolored bones may be bleached in diluted hydrogen peroxide and water with a trace of household ammonia.

Owl pellets will often reveal parts of the skeletons of small animals such as mice, shrews, young rabbits, a mole or small bird. Pellets of the kingfisher and herons will have fish

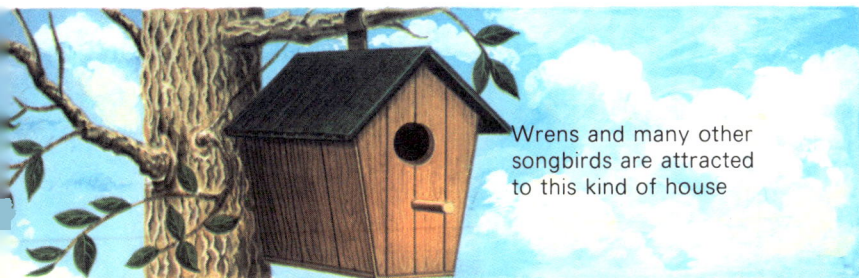

Wrens and many other songbirds are attracted to this kind of house

Feathers can be made into interesting collections.

skeletal remains as well as those of frogs and an occasional small mammal. Gulls and some of the waders may also produce pellets, but the typical one is ejected by a bird feeding upon mammals and the presence of hair helps to bind the mass together.

Untreated pellets and the bones may be mounted or put away in clear tubes or vials or even stored in envelopes. In some cases, entire skeletons can be reconstructed. Be sure to include all information known about where the pellet was found, what were the contents and, if possible, what bird it was from.

Individual Birds

So far, emphasis has been on the activities in which one might engage with regard to birds. Since there are 1,780 species of birds living or breeding in North America, it would be impossible to mention all of them in this book. However, the book is not complete without mention of some of the more common groups. Individuals have been chosen as representative of each group. Here, too, there are many which go unmentioned due to the space limitations.

Birds of Prey

Hawks, eagles, ospreys, falcons and owls are predatory birds. They are characterized by unusually strong, sharply clawed, grasping feet and heavy, curved beaks which enable them to tear their victims apart.

Bald Eagle: The American Bald Eagle, national emblem of the United States, is an endangered species due to the ways of man. Once plentiful, these majestic birds have been killed needlessly and so many of their nest sites destroyed that they are now rarely seen.

Sparrowhawk, or Kestrel: Smallest and most common falcon in North America, this bird frequents open and semi-open country and may often be seen perched on telephone wires or poles from which it hunts. It is the only small hawk which frequently hovers. Despite its common name, these birds eat primarily insects and mice although sparrows do make up the majority of birds taken as prey. Red-brown tails banded with black, chestnut-barred backs, an ashy-blue crown and black patches on the side of the head are identifying features.

Peregrine Falcon: These fierce birds of prey attack sea birds, ducks and pigeons. They may be seen over cities or towns. Slate-gray backs with darker bands, white under-sides with a buffy tinge to the breast and a heavy black 'moustache' line on top of the head mark this bird.

Barn Owl: These often appear when in flight in the dusk to be all white, but the plumage is actually sandy colored with a few white dots, the underparts being white with a few gray specks. They live in towers, barns or farm buildings, where their pellets can often be found.

An owl pellet, from which bones have been obtained

Golden Eagle

Peregrine
Falcon

Scavengers

These birds eat a variety of food material—insects, amphibians, young mammals, carrion, and even corn. Except for the vulture, whose hooked beak indicates its almost pure carrion diet, the others have long, stout beaks. All have weak feet, useful only for walking and perching.

Turkey Vulture: These common carrion-eaters often seen scavenging along roadsides are larger than the hawks with a wingspread of up to six feet. Vultures have a characteristic flight, tilting from side to side as they soar in lazy circles for hours on end. Their wings form a broad "V", and their folded-back necks give them a headless look.

Jays, Magpies and Crows: These fairly large birds are, for the most part, bold and aggressive. They are known for their raucous voices and large appetites. Their diet is varied, ranging from seeds and nuts to carrion.

Crows, solid black in color, can be found practically anywhere in North America although they are most abundant in farm areas. They are miniature versions of the raven of the far north and west which is nearly two feet in size.

The Long-tailed Magpie is restricted to the western half of the continent where, if in large numbers, it may be considered a pest.

Although there are numerous jays in the west, it is the Blue Jay of the east which is best known. This attractive crested bird is often considered a pest at feeding stations, but its loud "jay, jay" often warns other birds of danger. Males and females are identically marked.

Blue Jay

Crow

Game Birds

These birds, hunted at specific seasons of the year for food or sport, include, along with those mentioned herein, many of the ducks and geese grouped with water birds.

Quail, Grouse and Pheasants: These chicken-like birds scratch for their food. Most of them congregate in family groups. Smallest in size are the stubby-tailed quail, best known of which is the Bob-white, a bird named for its call. Other types of quail inhabit more restricted ranges in ·portions of the western United States.

Grouse are medium-sized birds having moderate to long tails depending upon the species. The majority of them have feathered legs which, especially in the case of the Ptarmigan, inhabitants of northern tundra and high mountain areas, enables them to survive winter weather. Also protective is their coloration which changes from brown in the summer to snow white in the winter. The Ruffed Grouse is known for its 'drumming' to attract a mate.

Originally native to China, Ring-necked Pheasants have been introduced to many parts of the world. They are now common in many parts of North America, particularly about farmland and in open woods where cover may be found. As with most pheasants, the brilliantly colored males outshine drab females. The white collar of the male Ring-neck is what gave them their name. These are probably most important of the game birds as they can be reared in captivity, yet quickly revert to the wild when released. Many states have extensive raising and releasing programs.

113

Ruffed
Grouse

Bob-white

Water Birds

There are many types of water birds. Some are oceanic birds, like gulls and terns, with long pointed wings that enable them to soar over ocean waters for long periods of time. Shore birds, such as sandpipers and plovers, usually feed along shores but may be found at inland marshes or grassy areas. Cranes, rails and coots are normally birds of marshy areas although coots often swim in open areas. Long-legged, long-necked waders are the herons, bitterns and egrets which may be found in fresh or salt water feeding on aquatic animal life. Perhaps the most familiar of water birds are the swimmers and divers—ducks and geese, swans, loons and grebes. Most of these have partially or fully webbed feet. Their beaks may be broad and flat for straining or sharply pointed for spearing. Short descriptions of some of the more common ones follow.

Herring Gull: This gull is abundant along coastal areas and yet common inland on lakes, rivers and garbage dumps. Although primarily a scavenger, this bold bird likes shellfish and has the intelligence to be able to open them. Unable to pry them open, gulls carry clams high above rocky or paved areas and let go. If the clam does not smash to bits upon landing, the gull will repeat the performance until it does.

Killdeer: A bird which repeats its name as a call, this is one of the plovers often found far from water, usually in fields and pastures. A ground nester, it often feigns injury to distract intruders. Adults have two black bands encircling their necks.

Coots: Often mistaken for ducks at first glance, coots congregate in large flocks on open water. They are differ-

Killdeer

Coot

entiated from ducks by their thick, high, white bill, a solid gray coloration and a habit of head nodding while swimming. Closer inspection reveals lobed toes rather than webbed feet.

Great Blue Heron: Largest of the darkly colored herons, these may be found wading in fresh-water or salt-water wet areas. They move stealthily through the shallows or stand, head hunched upon shoulders, watching for fishes which they catch with a quick thrust of their sharp beaks.

Swans: All white and largest of the North American waterfowl, swans have necks which are longer than the length of their bodies. This enables them to feed on aquatic vegetation without diving. Even though they are graceful upon the water and in flight, they lumber along on land.

Geese: Truly the middlemen of waterfowl, geese are much heavier and longer

Great Blue Heron

Canada Goose

115

Pileated
Woodpecker

Downy
Woodpecker

Ruby-throated
Hummingbird

necked than ducks, but not nearly the size of swans. Most common, Canada Geese, with a black neck and a head with white cheeks, often will flock in family groups. The Brants look like miniatures without the cheek patches. In some parts of the country, a mixed flock of Blue Geese and Snow Geese is a common sight.

Ducks: There are many types of ducks from tree ducks that graze, goose-like, to mergansers which have a specialized bill for an all-fish diet. Redheads and Scaup are some that dive for plant or animal food while there are others which merely dip their heads under water. Best known of these surface feeders, the Mallard, with its loud quack thought of as the typical duck-voice the world over, is one of those found in park ponds. Not all ducks quack though—they may whistle, moan, peep, croak or cluck. Many ducks are highly regarded as game birds.

Shoveller

Woodduck

Mallard

Redhead

Green-winged Teal

Pintail

117

Unusual Birds

Woodpeckers: Special adaptations enabling woodpeckers to probe beneath bark for insects include a strong, chisel-like beak, extensible barbed tongue, stiff tail feathers used as a prop when feeding on tree trunks and, a two-forward, two-backward toe pattern which also aids in clinging. Woodpeckers range from the sparrow-sized Common Downy to the uncommon crow-sized Pileated and Ivory-billed Woodpeckers. Flickers forage on the ground in search of ants.

Hummingbirds: Smallest of North American birds, hummingbirds have long, slender bills and tubular tongues for sipping nectar from flowers. Enabling them to hover in front of blossoms are wings which beat so rapidly that they produce a humming noise. Unlike other birds, they can also fly backward. Most species have iridescent feathers giving them a jewel-like appearance.

Perching Birds, or Songbirds

Perching birds, some 300 species, all have three toes in front with one behind, making them ideal for perching. There is, however, a great variety in size, coloration, habitat, habits and body structure. Beak structure is especially important to notice. Birds with heavy, conical beaks such as cardinals, sparrows and grosbeaks are seed-eaters. A long, thin beak, such as wrens, nuthatches, warblers, blackbirds and orioles have, indicate insectivorous birds. The heavier-billed tanagers and waxwings are mainly fruit and berry eaters. Short, wide beaks are typical of flying insectivores such as swallows, swifts and flycatchers.

Swallow

House Sparrow

Yellow Warbler

Black-capped Chickadee

Cardinal

Baltimore Oriole

Rose-breasted Grosbeak

Red-winged Blackbird

119

Mammals

Man belongs to the most highly developed of animal groups — the mammals. Although fishes rule the waterways and birds are supreme in the air, mammals are dominant on land. Even though mammals have certain characteristics in common, there is great variety in shape, size and way of life. For convenience, zoologists have classified the more than 3,500 species of mammals into smaller groups on the basis of physical characteristics. We shall look at some members of these groups in the following pages.

Marsupials

Opossum: The Opossum is the only North American pouched mammal. In this group, young are born before they are completely developed. They find their way to the mother's brood pouch where they are nourished and protected until able to exist outside. Frequently, young use their scaly, prehensile tail to cling to the mother's tail as they are carried around when they get older. An adult Opossum is a cat-sized creature which is most active at night when it hunts for small mammals and birds, eggs, insects and fruit. Threatened, it 'plays possum,' appearing to be dead.

Opossum

Bats often rest upside down, clinging to the ceilings and walls of deserted buildings.

Flying Mammals

Bats are the only mammals which can truly fly. As most bats are active at night, dawn or dusk, they are not very familiar creatures and many misconceptions and superstitions exist regarding them. Many of the 180 North American species of these insect-eaters are tropical in habit.

Most common are the Little and Big Brown Bats which are often found around human habitations. On summer nights, they may be seen flitting about lights feeding on the insects attracted to them or skimming pond and pool surfaces quenching their thirst. During the winter, they hibernate deep in caves or in the warmth of attics. A migrating species, the Red Bat, spends summers in the north and winters in the south. These chestnut or orange-red colored bats dangle from tree branches looking like faded leaves. Other United States bats are Leaf-nosed Bats, Pipistrelles, Hoary, Yellow and Silver-haired Bats and the guano-producing, Free-tailed Bat which led to the discovery of Carlsbad Caverns by its mass exodus each night. Bats provide valuable services as insect destroyers and should be protected.

121

A Pipistrelle Bat with its wings spread out reveals the modified structure of a five-fingered hand which forms its wing.

Insectivores

Moles: Small, muscular creatures with dark silky coats, moles have long snouts and, although their eyes are hard to find, they are present. Moles live nearly their entire life beneath the surface, digging with their strong foreclaws. Voracious feeders, a captive mole will eat 60 worms a day. They construct underground chambers which may be used for years. A unique mole is the Star-nosed Mole which has a cluster of fleshy tentacles on its snout. These help it to find the aquatic insects, crustaceans and fishes which it devours as it swims about.

Shrews: These are much smaller than moles and look more like a mouse with a longer tail. They emit a strange musky odor. Also voracious feeders, shrews eat four times their body weight in a day. They die of fright easily, especially when caught in a trap. Shrews feed on insects, worms, small rodents and snails. The five main varieties in North America are the Common, Pigmy, Water, Least and Short-tailed Shrews. Short-tailed Shrews are unique in that they are venomous.

Shrews, for their size, are among the most aggressive of mammals.

A mole is equipped with spade-like front feet which aid it in burrowing.

Cottontail Rabbits are a basic link in many foodchains.

Rabbits and Hares

Cottontail Rabbit: Found throughout eastern North America wherever there is sufficient food and cover, Cottontails are even found in suburban gardens or parks. In some cases, the decline of natural predators such as fox, owls and hawks has led to problems of rabbit overpopulation—a female may have three to five litters a year of up to eight young at a time and the young are mature and ready to mate at five months! Rabbits can be seen feeding during twilight hours along road edges.

Snowshoe Hares: Found in the northern states, these animals, like the weasel which may prey upon them, change from a brown summer color to white in the winter. This gives their other name of Varying Hare. Big feet, covered with long, coarse hair, act as snowshoes; thick fur and rather short ears, cutting down on heat loss, enable them to survive winter harshness. These animals are one of the primary foods of the lynx and coyote during the long winter.

The Snowshoe Hare changes color with the season—they are white in winter, then gradually turn brownish.

Foxes are chiefly nocturnal animals, except when disturbed during the day.

Carnivores

Red Fox: Possibly the world's greatest destroyer of mice is this dog-like creature. It will also eat other small mammals, insects, carrion, vegetables and fruit, game and poultry. The latter food choices often led to unnecessary slaughter. Foxes are also taken for their pelts, utilized widely by furriers. In addition to the reddish coat which is so familiar, there are blacks, silvers and yellow-browns. Their relative, the Coyote, of the central and western states, is better known for its voice than its fur. Its high-pitched howling during evening hours is very disconcerting to a stranger in its territory.

Weasel: Another important controller of the mouse population are the weasels, well formed for invading the tunnels of burrowing animals. These slinky creatures are bold and very aggressive. Residents of northern areas turn white during winter months while retaining black tail tips. Their valuable pelts are known as ermine. Even more valuable fur-bearers are the minks which can be raised in captivity.

Weasels are mainly nocturnal, but sometimes active by day.

The Harbor Seal can be seen near the coast from California and the Carolinas northward.

Harbor Seal: These aquatic animals, with limbs modified into flippers, are too wary and solitary to be seen often even though they inhabit harbors and the mouths of rivers and bays. This is the only seal which lives along the inhabited North American coast. It has escaped slaughter as it does not congregate in large rookeries like the Alaskan Fur Seal. Much larger than the 200-pound Harbor Seal, the Sea Lion of the Pacific coast may weigh 1,500 pounds and be 12 feet long, yet it is equally as graceful in water.

Striped Skunk: One of six species in America, all of which are black with white markings, the Striped Skunk ranges over the entire United States. Having a bad name because of its odor, a protective device released only when frightened or hit by a car, the skunk is actually a very beneficial animal. It is an efficient mousetrap and eats many garden insect pests as well as being important as a furbearer—Alaska sable and marten are actually commercial names for skunk fur!

Skunks have an unusual defensive mechanism—an unpleasant musky spray.

Bears: Weighing nearly three quarters of a ton, Alaskan Brown Bears are the largest living carnivores in the world. Because of their large size, they do not depend solely upon meat but will eat other foods—berries, nut, fruits, roots, and even grass. Their real favorite though is the salmon which they easily catch during spawning seasons. Fortunately for man, these gigantic creatures inhabit sparsely populated areas along the western coasts of Canada and Alaska.

Smallest of the bears, found throughout nearly all of North America is the Black Bear, which may also be brown, reddish-cinnamon, blond, or white. Though primarily forest dwellers, these bears are often found scavenging in local dumps. They have voracious appetites and are very fond of ant eggs and sweets such as honey and sugar. Campers must be very careful not to store food in their tents when they are in bear country. These bears are excellent swimmers and climbers even though, like other bears, they are quite nearsighted.

Largest of the bears found south of Canada, the Grizzly is extremely strong and ferocious. Despite weights of up to 1,000 pounds, these clumsy-looking animals can run as fast as a horse. A blow from their mighty paws may break the neck of a bison. Formerly found throughout western North America, they are now limited primarily to park areas.

More commonly seen in zoos than the wild, the Polar Bear is the only aquatic bear. Its white coat suggests its icy habitat —ice floes and snowy Arctic coastal areas. These home

Polar Bears often haul up on Arctic ice floes.

territories limit its diet somewhat to seals, walruses, fish, crustaceans and occasional stranded whales, but it will even eat seaweed if hunting is poor.

Hoofed Mammals

Deer — animals with antlers: The deer family of North America is widely diversified: it ranges from the massive Moose, a 7-foot-tall creature weighing up to 1,400 pounds to the rare Key Deer, a pygmy at 30 pounds with a shoulder height of 2 feet. Most common is the White-tailed Deer which ranges over most of the United States and southern Canada. This deer is one of the few animals which has benefited from man's spreading habitation — it prefers an open woodland, forest edge, or pasture to dense woods. At times it may become a pest in orchards.

The Mule Deer is common on western ranges. Outstanding are its large, mule-like ears. It usually has a dark forehead, white rump patch and black tail tip. A sub-species of the Pacific coastal region is the Black-tailed Deer.

Largest of the round-antlered deer is the Wapiti, or Elk. Bull Elks may weigh 1,000 pounds and have antlers that span 5 feet. Spotted young weigh 30 pounds at birth.

Most awkward appearing of the deer is the Moose with its long legs, humped shoulder and strange muzzle. It is hard to believe that the 60-pound, 6-foot, spreading, flattened antlers of the bull are shed annually.

Black Bears are quite common, but the Grizzly Bear is an endangered species.

White-tailed Deer flash a white
rump patch when alarmed.

Elk are gregarious mammals, often
seen in large herds.

Moose feed frequently on rooted
aquatic plants which they pull
up from shallow ponds.

Pronghorns, once nearly extinct, are now abundant in many western states.

Pronghorn Antelope: Found in open country and plains areas from Canada through northern Mexico, the Pronghorn is the swiftest of American mammals. It can run at speeds of between 40 and 50 miles an hour. Formerly numbering in the millions, the present population is much less. Although it can make tremendous horizontal leaps, fencing has limited its range tremendously as it cannot jump vertically. In bad winters, hundreds may starve to death within sight of fenced-in haystacks.

Bison, or Buffalo: Originally ranging over most of the United States, these huge animals almost became extinct and can now be found only in parks and preserves. Most gregarious of all wild cattle, these creatures gather in large herds which facilitated the mass slaughters at the turn of the century. Bulls may weigh up to a ton.

Bison once provided food, shelter, clothing, and tools and utensils for many Indian groups.

Rodents

Gray Squirrel: This woodland and park dweller may be the only wild animal city children are familiar with. Often hand-fed and seemingly tame, they may bite severely when frightened or handled. They are useful as nut tree planters—many acorns and nuts buried are not relocated!

Chipmunk: A small, squirrel-like animal with a flattened hairy tail, the Chipmunk is best known by the five black and two white stripes running along the back and sides of his chestnut body. He also has a stripe from the tip of his nose past his bright black eyes. This and a smaller size distinquish him from his larger cousin, the Golden-mantled Ground Squirrel found in the Western states.

House Mouse: Found wherever man dwells, the House Mouse may cause a great deal of damage. Hybrid mice are fairly commonly kept as pets. There are many varieties of color and size, including those with hereditary peculiarities such as singing mice and waltzing mice. A close relative, the White-footed Mouse is frequently used for experimental purposes, as it is quite susceptible to human disease.

Norway Rat: Even though albino forms are useful laboratory animals, these are our worst pests. They cause a great deal of agricultural damage, and carry much disease. These rats emigrated to North America with the colonists. They are extremely prolific, having about eight young in each of as many as six annual litters, . . . the young are ready to bred before they arc six months old!

Woodchuck: Commonly seen along road edges in northeastern states, Woodchucks are well-known for their hibernating habits. In fact, Ground Hog Day is a much-celebrated holiday in some areas. Woodchucks, or Ground Hogs, are practically pure vegetarians. They are especially fond of young garden plants and may even climb over low fences to get into home gardens. Tiny ears, thick fur, whiskers as long as their body is wide (to prevent them from going into an area where they might get stuck) and strongly clawed digging feet indicate their burrowing life habits. Marmots of the west are close relatives.

Gray Squirrel

Chipmunk

White-footed Mouse

Black Rat

131

A Woodchuck burrow may be 20 to 40 feet long.

Porcupine: The Porcupine is a slow-moving, clumsy rodent. It is found in most of the northern states and in Canada. Outstanding is this animal's protective device, a coat with some hairs specialized into barbed quills. Once a predator has gotten an noseful of quills, he is not likely to attack another porcupine. Porcupines feed on twigs and bark and may become serious pests in pine forests.

Beaver: Largest of all North American rodents, the 40-pound Beaver has such a valuable hide that it led to the exploration of the Rocky Mountain area. The rodent's chief characteristic is its oversized incisors—teeth which serve as both tools and weapons. Beavers are quite well known for their engineering skills. Using their sharp incisors, they cut down young trees which they then trim, using both log and branches to build a dam in order to form a pond. The pond serves a two-fold purpose: it gives them a place to build their lodges with their underwater entrances and it provides a place for food storage. A side benefit to man is the part dams play in flood control.

When alarmed, Porcupines lash with their tails, driving quills into anything with which they come in contact.

Beavers dam up streams to form the ponds in which they build their lodges.

Things To Do With Mammals
Field Study

Since most wild mammals are secretive and wary, it is difficult to observe them close-up in their natural habitat. To do so takes skill and infinite patience. A good pair of binoculars helps bring the animal much closer to you when you must keep at a distance. Telephoto lenses on cameras also help both in photographing and observing. Powerful flashlights aid in observing nocturnal animals. Prepare by scouting an area in advance to pinpoint the haunts of an animal—look for signs of its home, feeding, watering, and resting areas. Staking out an observation post near one of these areas is likely to be more fruitful than haphazardly wandering about. In some cases, bait or calling will help attract animals. This is particularly helpful to the wildlife photographer.

Preparation of a Skeleton of a Small Mammal

A rabbit will serve as a useful example. First the animal is skinned and the gut contents removed. The animal is then fleshed as much as possible without cutting any of the ligaments which hold the joints together. The specimen is then placed in a diluted washing soda solution, say a teaspoonful to a small saucepan of water, and gently boiled. The flesh soon comes away from the bones, and the specimen should then be removed from the solution and washed under cold

133

tap water. Any meat still attached to the bones can be removed by making a thin cream from household bleach and brushing it gently over the skeleton using an old nylon toothbrush; bristle brushes are of no use as the bristle will dissolve in the bleach. If necessary, the head can be removed first, then the forelimbs, hind limbs and other parts processed separately. However, it is possible to prepare a small mammal of this size without separating the skeleton at all.

Preparation of a Skin of a Mammal or Bird
This work takes a long time, and only experience will give good results. It is a good plan to practice on a rabbit first and learn how to remove the skin without damaging it. It is possible to remove the skin of the feet by rolling it off like a sock, leaving the skeleton behind. The same method can be used for the head, except for the lips and end of nose which must be cut through. When skinning birds, it is helpful to remember that the beak is attached to the skull and should be left in the skin. A piece is usually taken out of the back of the skull and the brain cleaned out, removing the eyes at the same time. Mixtures of salt and alum are used to dress the skin. Sometimes ashes are used. Professionals use arsenic and other poisonous compounds, which should not be used by the amateur. Common salt can be used successfully by learners in this field.

A skeleton of a rabbit

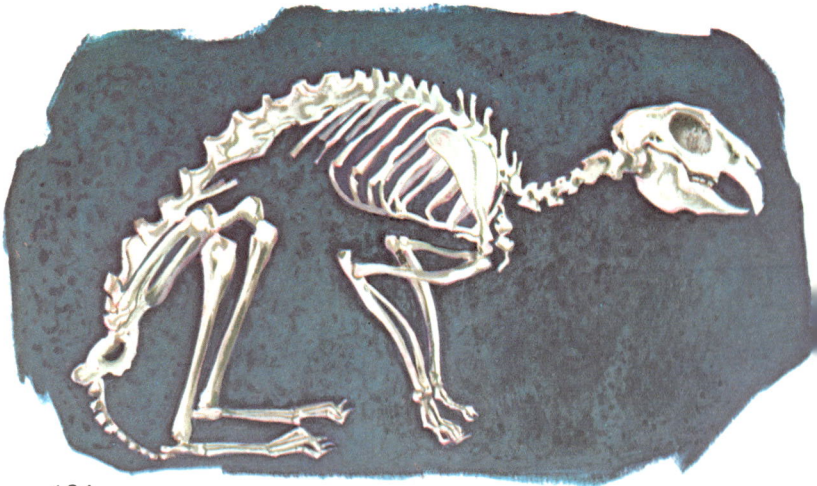

Plaster casts of a deer hoofprint

Positive

Negative

Plaster Casts

An interesting way to collect and keep records of footprints of various animals, leaves, or plants is to make plaster casts of them. Dental plaster is used for this purpose and is readily available from druggists. Add the powder to water to make a paste with the consistency of condensed milk. Make a low plasticine or cardboard wall around the impression, pour in the plaster and allow it to dry for at least 20 minutes.

The mixture will get warm as it dries, and only when it is quite cool should the mold be moved. This will give a negative impression; the positive cast of the footprint or leaf can be obtained by pouring more plaster mixture over the hardened mold after greasing it well.

Keeping Mammals As Pets

Man has kept some mammals as pets for thousands and thousands of years. Domesticated animals such as horses, dogs and cats are often raised primarily for this purpose. However, capturing wild animals to be kept as pets places the animal in an unnatural situation. In many states, it is illegal to trap animals out of the regular hunting season. Often special permits are needed before wild animals may be kept. It is best to check state and local regulations before keeping or trapping any wild animal. If permitted, you must then be sure that the animal can safely be trapped, caged, and well maintained.

A knowledge of natural habitat, habits and food preference of the animal to be kept is essential for providing proper housing and care. Caging selected should be large enough to allow for free movement, reproduction, and if possible, a portion of the natural environment. It should also be easy to clean.

Food and water containers should always be kept clean and a fresh supply given daily. Containers should be shallow enough for the animals to feed without having to climb in them. They also should be fastened or constructed so that they cannot be tipped over. Water bottles are ideal for most mammal cages.

Outdoor cages are best for native mammals for all but the coldest weather. Sleeping quarters should be raised off the ground.

When handling wild animals, it is important to remember that most animals will bite or try to bite if frightened. Even though they may seem tamed, there may be times when they react adversely to actions, noises, or some stimulus which we might not even be able to discern. It is best to wear heavy gloves when picking up strange animals. Remember, too, that only young animals tame readily; some adults may never be tamed. Patience, gentleness and quiet, slow movements are all necessary in the taming of any animal.

136

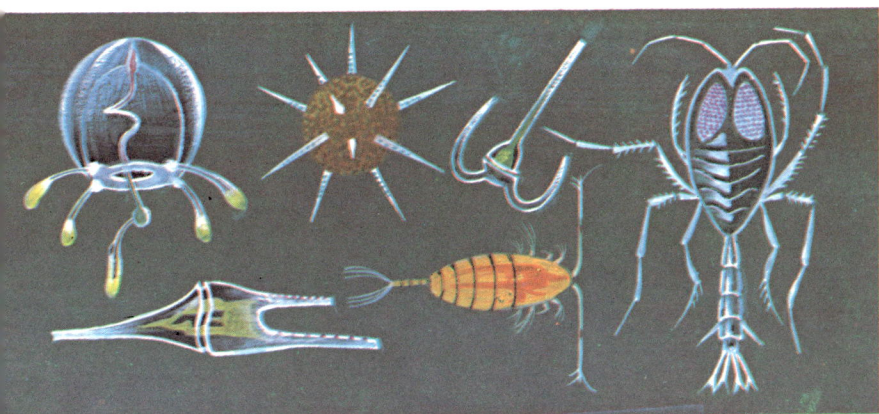

Minute Organisms

To end the chapter on plants and animals a short note follows on the plankton, the drifting plant and animal life found at or near the surface of the sea and in freshwater lakes and rivers.

These are mostly small organisms and are in many cases the young stages of many of the invertebrate animals, although some vertebrate forms are found together with many primitive plants, mainly algae. Practically all these organisms are transparent, and their beauty and fascination has to be seen to be believed.

Fine nets are used to collect the samples, which are then placed in a jar with a few drops of formaldehyde. This kills the organisms, which fall to the bottom of the jar. Drain the fluid and replace with a fresh solution of 5 percent formaldehyde. Pencil-written labels may be placed inside the jar but if a dense sample obscures vision, an outside label may be needed.

All of these minute organisms are plankton.

COLLECTION AREAS

Grassland

When the explorers and settlers first came to America, they found a continent which was just about half forest and half grassland. At that time, the grasslands abounded with many grazing mammals—buffalo, antelope, deer, and elk. As man invaded with his domestic grazers, the native animals were driven out. In addition, farming ruined the sod. Over-grazing and poor farming practices led to the destruction of the grassland. Increased populations of burrowing animals, mice, gophers and prairie dogs, and plagues of grasshoppers added to the problems. Poisoning campaigns to cut down the rodent population also killed off foxes, coyotes and hawks, the natural predators. Fortunately, much has been done to restore grassland areas. Fireweed, goldenrod, daisies, and sunflowers are again attracting insects which, in turn, attract insect-eating birds. In this way the balance is being restored.

Marshes are a special type of grassland—a wet, treeless land characterized by grasses, hedges and cattails, just as a swamp is any wet area dominated by trees. There are two different kinds of marsh—salt marshes, within the influence of tidal waters, and freshwater marshes. Marshes are rich in vegetation and wildlife—muskrats, rails, bitterns, coots, ducks, warblers, red wings, turtles, frogs, snails, fish, crayfish, dragonflies, and water beetles are some of the animals which may be found there. Flowering plants such as gentians, marsh marigolds, and shooting stars may also be seen.

Nearly destroyed by misuse, grasslands have now been restored in many areas, and endangered species such as prairie dogs and coyotes have become reestablished.

Forest

The definition of a forest as a treed area is misleading as there are many types of forests. Variation in climate, elevation, soil condition, rainfall and sunlight exposure determine the kind of forest in any area. Northern forests of spruce and pine are inhabited by squirrels, porcupines, jays, owls and small tree-nesting birds while twinflower and pyrola are found blooming on the forest floor.

Eastern deciduous forests may be primarily maple, oak, beech or birch or combinations of these with some evergreens mixed in. These too, have a great many tree-nesting birds, flying squirrels, deer, rabbits, opossums, raccoons, mice, snakes and amphibians. There is a wide variety of shrubs and forest floor plants. Characteristic understory shrubs include the sassafras, witch hazel, mountain laurel and dogwood. A thick humus on the forest floor provides nutrition for fungus such as puffballs, boletes and coral fungus, lichens like the British Soldiers, and a wide variety of mosses and ferns plus homes for beetles, worms and toads. Flowering plants include lady slippers, trilliums, wild geraniums, Jack-in-the-pulpits and skunk cabbage.

Tropical forests and rain forests have an abundance of epiphytes such as orchids and hanging mosses, many amphibians, insects (especially mosquitoes), ferns and mosses. While northern rain forests have redwoods, cedars and huge oaks, the tropical forest is characterized by cabbage and thatch palms, mangroves and mahogany.

Forest plant and animal species vary with the kinds of trees which predominate in a particular place.

In fresh water, the kinds of plants and animals one can observe depends on whether the water is standing or running.

Fresh Water

In still ponds and lakes there will be a profusion of water plants, from the lowly bacteria, invisible to the eye, to the beautiful iris or flag. Many aquatic plants have special adaptations, one very interesting example being the bladderwort which captures and digests living animals in small bladderlike tissues in its strangely modified leaves. The hydra, a member of the Cnidaria, is often quite common, although sometimes it disappears without a trace and does not appear again for many seasons. Water fleas appear in great numbers in fresh water, together with many snails, shrimps, and small mussels. There are countless insect larvae, especially those of beetles that pass their early life in the water.

Toads, frogs and newts are found anywhere there is a small natural body of fresh water. Water snakes, water shrews, rats, muskrats, otters and beavers may be found. In fact, the beaver may create a pond from a small trickle of running water. Once plants and insects have become established in a young pond, fish are able to survive—killifish, sunfish, minnows and carp.

There is a wider variety of fish in running water—catfish, trout, bass, and the lamprey with its curious sucking mouth. Only the insects which can find protected areas or have developed special devices to prevent them from being swept away by the current can be found. Some of these are water pennies, stoneflies, and caddisflies, with their strange larval cases and nets they make for trapping food.

140

Seashore life includes many birds which feed on the small fish, crustaceans, jellyfish, limpets and other species which live there.

The Seashore

The life to be found on seashores is as varied as the many kinds of seashores that exist. Rocky shores contain bare rocks, weed-covered areas, rock pools and large boulders which may conceal small pools. These rocky areas shelter animals and plants capable of clinging to rocks. Limpets and sea weeds will be found, and on the weed, many periwinkles and small dog whelks. The occasional chiton is found on exposed rocks, and there will be many barnacles in the areas well washed by the sea, together with masses of mussels clinging by the silky byssal threads. The rock pools contain a wealth of plant and animal life. Algae are well represented, and attached to them are many of the colonial hydroids and sponges and the odd bryozoan covering the surface of the algae with its peculiar lattice-work skeleton.

Shingle beaches are conspicuous by the absence of any obvious animal life, although below tide level there are many creatures that live in crevices between the pebbles and stones. On sandy shores there is little or no plant life to be seen, but by digging, a profusion of animal life will be found. Burrowing worms, crabs, mollusks and many other animals will be found. Muddy shores also have their own peculiar burrowing animals, especially when clay is present.

On any beach with pools left by the tide there are many fish and small crustaceans; jellyfish may be found and anemones are often present. Gulls, terns and shore birds abound where food and shelter can be found.

Desert

Many people think of a desert as a barren, lifeless place. Life in these extremely dry, often hot, areas is not abundant in comparison to other habitats, but it is quite numerous considering the harsh environmental conditions. Desert denizens range from snakes, lizards and insects through birds to mice and kangaroo rats, cottontails and jack rabbits, ground squirrels, badgers, skunks, pumas, bobcats and coyotes. Animal life of a particular area is dependent on the water supply, food and shelter available. As such, it is closely related to the type of vegetation which may range from spiny cactus and creosote bush to live oaks and evergreens. Again, the plants which grow depend on water supply, soil condition and, in addition, the elevation, for contrary to popular belief desert areas may be mountainous as well as flat lowlands.

In the very arid areas with little vegetation, animal life is limited primarily to such lizards as the horned lizard; the spadefoot toad which is equipped to bury itself beneath dry desert sands for long periods, emerging during infrequent periods of rainfall to feed and breed; sidewinders which feed upon the kangaroo rats, mammals with special abilities to manufacture their own water; and some birds of prey which are able to fly into these areas to feed. In other areas, with more vegetation, desert tortoises abound along with many vegetarian lizards, quail, rabbits and all the predators which feed upon these.

Although deserts may seem barren, lifeless places, they support a large and varied group of plants and animals, among them the Kit Fox and the Kangaroo Rat.

142

COLLECTING FOSSILS, ROCKS AND MINERALS

Fossils

Fossils are the hardened remains of ancient plant and animal parts. They may have been preserved in a variety of ways.

Actual preservation is rare, but there are examples of animals that have been preserved for many thousands of years. These include the mammoth found in the tundra of Siberia, and the woolly rhinoceros found in oil seeps in East Germany. Many fossils are produced by *mineralization*. This occurs when porous shells and bones are filled with mineral matter. It is a protective action and the remains are usually in good condition. *Replacement* is a process that also produces good fossils. In this case the actual remains have long disappeared and been replaced exactly by silica or quartz, or perhaps by calcium carbonate or iron pyrites. Fossil plants are often preserved by *carbonization*. In this process, all the plant tissue has decomposed, leaving a residue of carbon to record the appearance of the actual organism. By *molding and casting,* shells are often dissolved, leaving a cavity or mold of the exact shape of the original in the rock. The fossilized sea urchins often found in the chalk are often flint 'casts' of the original animals. *Tracks and imprints* of animals are often preserved in mud which has hardened into rock. Many imprints of invertebrates have been found, especially of jellyfish and worm burrows. *Coprolites* are examples of fossilized excrement or dung, in which scales of fish are sometimes found and other parts of devoured plants and animals. *Encrustation* preserves remains when water in caves in limestone rocks drips on to objects. The water contains minerals that form around the remains.

Various conditions favor the preservation of fossils.

Fossil insect and (*right*) starfish

Rapid burial in sediments or muds is one condition, otherwise dead animals and plants are devoured by scavengers or decayed by bacteria or enzyme action. Early burial in moist sediments prevents this action in most cases. Fine sediments produce better fossils than coarse sediments.

Rapid burial in volcanic ashes is also favorable to preservation, and a good deal of dinosaur material has been preserved in this way. A more modern example of this occurred

A collection of fossils

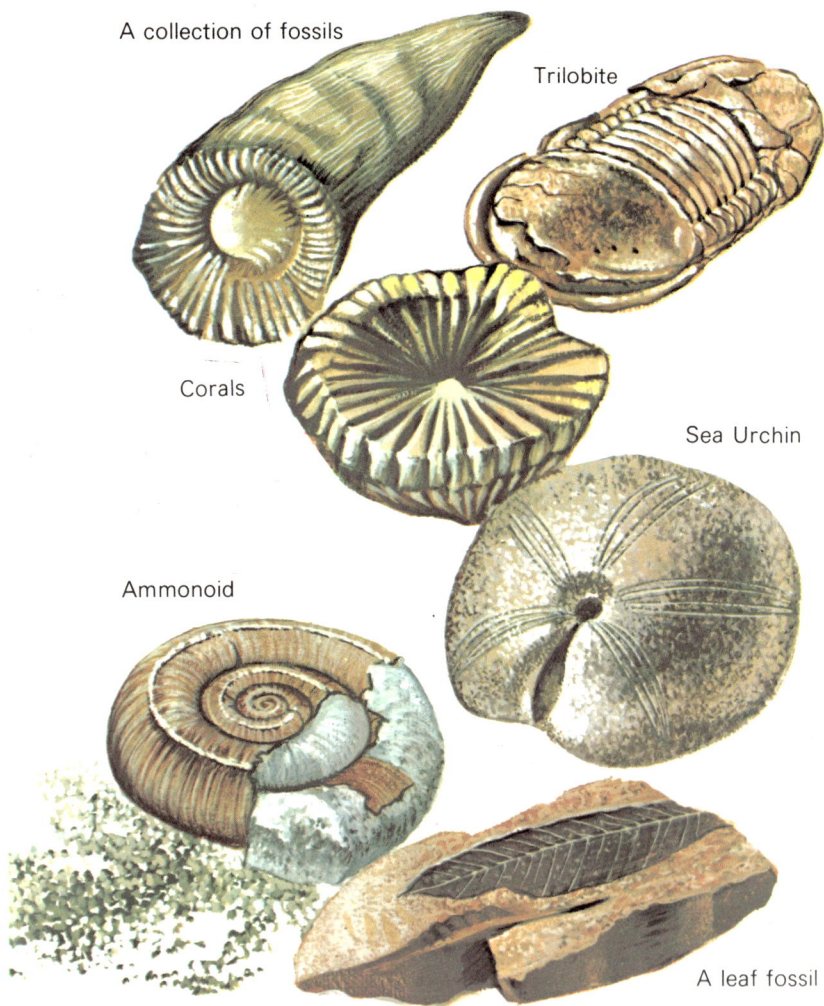

Trilobite

Corals

Sea Urchin

Ammonoid

A leaf fossil

Calcite

Actinolite

when Pompeii and Herculaneum were overwhelmed by Vesuvius in 79 A.D. Many people were literally fossilized overnight, as were their animals and plants.

The existence of hard parts is usually essential for fossilization. Many of the invertebrates with a hard outer shell, such as crabs, are well preserved, as are bones and other hard substances. A uniform temperature with no freezing or other extremes, and the presence, already mentioned, of waters of a high mineral content are also conditions favorable to fossilization. Conditions of quiet when deposition is taking place are also very necessary to prevent the animal or plant parts from being separated. Wave action and high winds would tend to spoil the remains.

Where to Find Fossils

Fossils are found in sedimentary rocks and it is the rate of the deposition of these sediments that helped to preserve the remains. Fossils are sometimes also found in lava or volcanic earth, and most commonly in volcanic ash.

Wherever road building is taking place, fossils may be found. Consult the geological map of an area to help in the search. Railroad cuttings are also good places to look, and canyon areas and quarries are good spots to begin investigations. Make sure that you have permission first.

The cliffs and the seashores are two other areas where fossils may be found, especially after storms in the winter, and any earth or rock-moving operation is likely to reveal fossil material.

Fossils are often weathered out of the rocks and can be

145

Iron Pyrites   Zincite

picked up quite easily. Often large flints brought to the surface in chalky areas by plowing may reveal sponge and echinoderm fossils when cracked open. These are often further revealed after a heavy rainstorm.

All fossils should be wrapped in newspaper and packed in plastic bags for transport.

Preservation of Fossils

Preservation consists mainly of repairing, hardening and in general strengthening the collected fossil. Small fragile fossils can usually be repaired with clear glue.

A solution of perspex chips dissolved in chloroform is one adhesive used by many professionals in the field. Gum arabic glue is also used, but it tends to help the growth of molds and a suitable germicide should be added to the solutions before use. A misture of plaster of Paris may be used to strengthen fossils in the rocks if they are fragile. Mix the plaster into a thick paste and pour it over the specimen.

Some form of controlled drying is useful for finishing off fossils collected from damp clays and gravels. A tin can over the gas burner serves quite well as an oven, but ordinary fossils will dry out in air. They may then be sprayed with a protective coat of clear plastic or varnish.

Rocks and Minerals

Minerals are usually collected during mining operations, and the old workings and dumps of abandoned mines are good sources of specimens. Care must be taken when looking

146

Colored pebbles, well polished, look extremely attractive

over these areas. Pebbles found on the beach are often a fruitful source of minerals.

Rocks are easily collected from cliff faces or wherever road building or other civil engineering is being carried out. Many good samples are weathered out in a similar fashion to fossils, by the effects of rain and frost.

Preservation of Rocks and Minerals

Apart from those rocks and minerals that are deliquescent (that become moist in the presence of air), the remainder may be air-dried, if necessary, and left at room temperature quite satisfactorily. Deliquescent rocks or minerals will need to be kept in a container containing a drying agent, such as

Small tumbler polishers can be easily purchased or made.

Polished pebbles

silica gel, which must be replaced from time to time to keep the specimen absolutely dry and well preserved.

Pebble Collecting

What is a pebble? Some say that pieces of rock from a minute stone no larger than a grain of sand up to a boulder are pebbles of one sort or another. The dictionary definition usually says 'a small water-rounded stone'. This raises the question of how much rounding is required to qualify as a pebble. It is probably most sensible to regard all stones on a beach as pebbles of various sizes. The same action of water is apparent in rivers and lakes, and the small stones found in these areas are *ipso facto* pebbles. The early and middle nineteenth century were the great days in pebble collecting, and many of the watering places, as they were called, around the coasts were the scene of intent people all carefully combing the beaches in search of the odd and unusual in these stones.

It is amazing how interesting a small collection of pebbles can be. It can produce great interest in the study of rocks from which they came, and the pebbles themselves are often of great beauty, color and shape. Often the mere mention of the word 'geology' makes people shy away from serious contemplation of these products of the earth's crust. Pebble collecting can be carried out on the beach, by the lakeside or by the river. Many of these pebbles are more attractive if they are kept in a jar of water.

Small pebbles may be polished by a process called 'tumbling'. A tin or similar container is set up so that it can be revolved slowly. A mixture of glycerine, water and carborundum powder is placed in the tin and the collected peb-

Agate

Schist

Gabbro

bles added. Slow tumbling over a period of weeks gives excellent polished pebbles. It is a fairly simple matter to assemble a piece of apparatus to do this. About four or five weeks in the tumbler are required to produce an excellent polish. Some of the smaller pieces can be used for making jewelry.

SETTING UP YOUR COLLECTION

Small collections may be kept in many places—perhaps a bench by a window in the school laboratory, or part of a bookshelf, or even a series of drawers in a cupboard or a series of boxes. Whatever form it takes, there must be order and method in the collection.

Whatever the objects collected, always get expert advice on the identification and treatment of them, and try to read as much as possible about your collection. Make sure that every specimen collected has an identification mark (a spot of paint or number) by which the labeling and description of the collection can be correctly carried out.

Labeling

A specimen whose origin is unknown is seldom of any value. It is of the utmost importance that all specimens be carefully labeled, and this should be done as soon as possible after collection. Labels should be placed *inside* the receptacles containing the specimens.

Information should be written on good quality rag-paper labels with soft pencil or in India ink. The collector should give the locality and position as accurately as possible. Examples of the necessary information include the nature of

Red Serpentine

Onyx Pebble

Granite

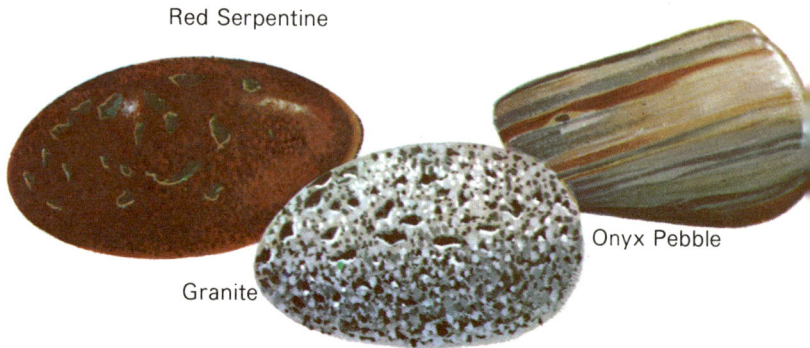

the habitat, the depth of the sea and type of sea floor, height above sea level when collecting on land, vegetation in the vicinity, date of collection and method of treatment of specimens collected. If the specimen is a parasite, the name and description of the host, and further information as to where the parasite was found must be given. The sex of animals should be recorded, and with plants, the type of soil where the specimens were obtained. A good label should be made out as follows:

1. Local name
2. Scientific name (if known)
3. Where found
4. Name of finder
5. Date
6. Anything else of interest

Displays

A haphazard collection has little value but a well-arranged collection can have real meaning. One does not need a lot of money and a lot of space to have an interesting display, but a little time and imagination will do wonders. Rather than just line up rock samples in a row on a shelf, try to display them in a way which will show how they are used by man, or perhaps, how they developed. This will be far more interesting, not only to your friends, but probably to yourself as you research such ideas. Nature displays can tell intriguing stories, not only about rocks and minerals, but about any of the natural wonders which you come into contact with daily, be they plant or animal.

Displays can be greatly aided with drawings, photographs, pictures from magazines, diagrams and models. Since it is so difficult to make stuffed specimens look lifelike and it is equally difficult to keep certain specimens from disintegrating with age, models of some specimens may be preferred.

Making Models

It is relatively easy for the amateur to learn how to make plaster of Paris or rubber molds of natural objects. From these molds, plaster, wax or rubber models or casts can be made which, painted and finished, can look like the original specimen. In many cases, they are much more durable and effective for display purposes.

Plaster of Paris mold impressions of leaves can easily be made.

The process is much the same as given for making casts of animal tracks, but may be more complicated depending upon what is selected as a model. The animal is placed in a form and prepared plaster of Paris is gently poured over and around it. If necessary limbs and the head may be supported with pins before the pouring commences. After the plaster set sets, the animal is removed and the mold must be thoroughly dried before being used to make your models. One of the best ways to do this seems to be to put it in a small cardboard box with the top open and a lighted 40-watt bulb near the top. Higher temperatures will cause the plaster to become powdery and break down. Once the mold is dried, it may be used with plaster, rubber or resin-wax mixtures to make the models. Directions are usually supplied with the compound which you use.

Mounting and Displaying Specimens

Rocks and minerals are usually either mounted on display boards or in compartmented trays or drawers. Each should have an identifying mark on it. This is usually done by painting on a white spot on which is written a number or the name of the specimen. Samples may be mounted on display boards with household cement or wired on. Small samples

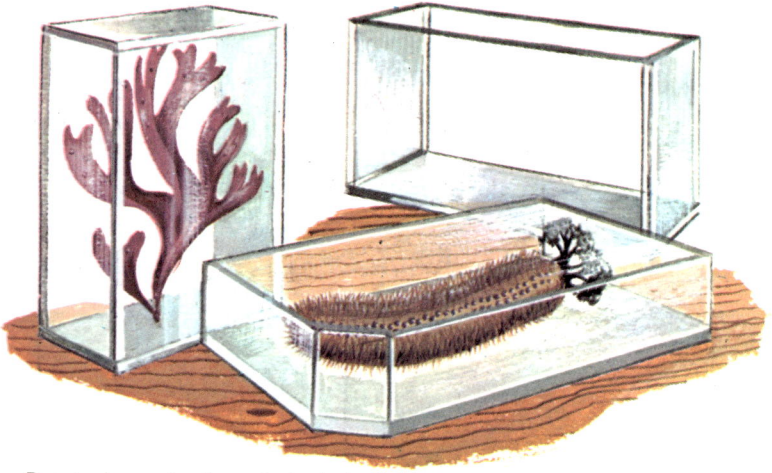

Preserving animals and plants in fluid

may be mounted in vials, particularly specimens of soil or chemicals refined from your various samples.

Many of the lower animals—worms, spiders, creatures of the seashore, and some insects—have soft bodies which are preserved in solutions of alcohol or formalin. These are usually kept in labeled jars or a display mount which can be made from a waterproofed frame and glass sides sealed with windshield sealer. Care must be taken when using the preservatives, as alcohol is highly flammable and formaldehyde is an irritant which should not be inhaled.

Natural history specimens are best stored in jars, glass-topped boxes or drawers.

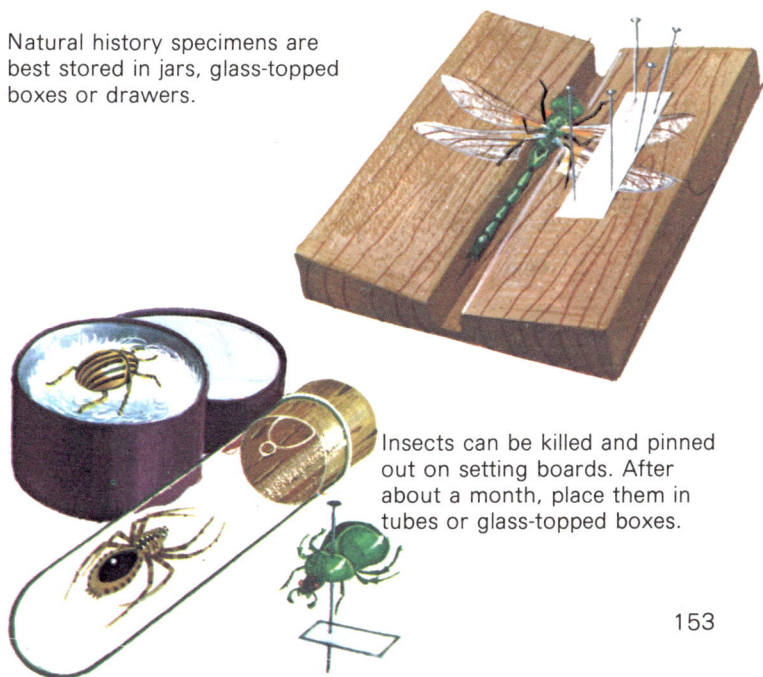

Insects can be killed and pinned out on setting boards. After about a month, place them in tubes or glass-topped boxes.

Caterpillars may be preserved by emptying the skin of the insides, easily done by rolling a pencil from head to tail, and then inflating the skin with hot air. These can be used to make displays of insect collections more interesting and informative by presenting a display of the life cycles of several insects. (For details regarding the preserving and mounting of the hard-bodied insects, see pages 16–17, 67–68.)

Higher animals, the birds and mammals, may be skinned, stuffed and mounted, but since this usually requires a bit of expertise, details will not be given. The bones, teeth, fur or feathers, portions of homes and samples of food of these animals may be mounted to form interesting exhibits.

Besides preserving plants as mentioned on pages 40–43, there are several other ways of preserving them or making reproductions of them which can be displayed nicely. One might mount freshly picked plants or leaves between a sheet of heavy paper or cardboard and a layer of clear plastic. Mounts of this sort, if sealed airtight, retain their natural color for long periods of time. Of course, this does not work with bulky plant material. A similar activity is the pressing of flowers or leaves between two sheets of waxed paper by using a warm iron.

Leaf prints may be made in a number of ways. Simplest of all is to press a leaf, vein side down, on an ink pad, a sheet of carbon paper, or a freshly painted surface. Then lift the leaf and carefully press it down on a clean sheet of paper. Smoke prints are very attractive and can be used for stationery as well. Coat a sheet of cardboard with wax or grease. Hold this, treated side down, over a flame, being careful not to ignite it. The cardboard will soot up. The leaf is then pressed into this, transferring the carbon to the vein side of the leaf. Following the same procedure as with the other types of prints, this is then transferred to a clean sheet of paper. This type of print gives a particularly delicate tracery of the leaf venation.

If you are only interested in an outline of the leaf, you may find spatter printing intriguing and easy as well. The simplest approach is to weigh down a leaf on a sheet of paper, dip a toothbrush in thick ink or poster paints and, holding the brush bristles up above the paper, rub a stick over the bristles. The resulting spatter leaves the leaf in white silhouette against a background of colored drops. Other similar approaches utilize a piece of screening held above the paper and brushed against, the use of atomizers or spray guns, or aerosol cans of paint. Other, more complicated methods include blueprinting, oxalid printing or making 'etchings' using plexiglass or lucite.

Twig and winter bud collections are simple to make and easy to mount for display purposes. Be sure to get specimens which are characteristic of the shrub or tree you are displaying. In the spring, bud collections make fascinating exhibits if the twigs are mounted in small tubes or vials of water. Collections can also be made of wood samples or bark sections. These can all be wired or secured with screws upon display boards.

Seed collections can be quite interesting. It is important to be sure that seeds collected are free of insects or your entire collection can be ravaged. Some seeds are best preserved by coating with shellac or dipping into a liquid plastic; fleshy fruits may have to be preserved in formaldehyde. Displays might show how seeds travel, might match seeds with their flowers, might show uses of them. Of course, wild seeds collected can also be used for starting your own gardens . . . a bit more of a challenge than the normal gardening.

These are just a few of the activities which may make your exhibits more interesting and their preparation more enjoyable for you. The number and type need be limited only by your time, imagination and the materials you collect.

BOOKS TO READ

Field Book of Nature Activities and Conservation. William Hillcourt. Putnam, 1961.

Field Book of Ponds and Streams. Ann Haven Morgan. Putnam, 1930.

Our Small Native Animals: Their Habits and Care. Robert Snedigar. Dover, 1963.

Tropical Freshwater Aquaria. George Cust and Peter Bird. A Grosset All-Color Guide. Grosset & Dunlap, 1971.

Bird Behavior. John Sparks. A Grosset All-Color Guide. Grosset & Dunlap, 1970.

Microscopes & Microscopic Life. Peter Healey. A Grosset All-Color Guide. Grosset & Dunlap, 1970.

The Insect Guide. Ralph B. Swain. Doubleday, 1948.

A Field Guide to Reptiles & Amphibians. Roger Conant. Houghton-Mifflin, 1958.

Ecology. Peter Farb. Life Nature Library. Time, 1963.

A Field Guide to the Mammals. W. H. Burt and R. P. Grossenheider. Houghton-Mifflin, 1964.

Pond Life. George K. Reid. Golden Press, 1967.

Non-Flowering Plants. F. Shuttleworth and H. S. Zim. Golden Press, 1967.

INDEX

Acorn worm 44, **44**, 74–75, **75**
Actinolite **145**
Aeration 84–85
Agate **149**
Alchemist 6, **6**
Algae **11**, 24, **25**, 26–28, **26–27**, 30, 85, 86, 141
Alligator 99–100, **100**, 102
Ammonite 11, **144**
Amphibians 77, **77**, 87–94, **87–94**, 139
Anemone 44, 51–53, **52**, 86, 141
Angiosperm 36, 38–39, **38–39**
Animals 9, 10, 14–19, 44–142, 151–154
  collecting 9, 10, 14–19
  fossils 143–144
Annelids 44, **45**, 60–62, **61–62**
Antelope 129, **129**, 138
Aquarium 83–86, **85**, 92
Aril (yew) 36, **37**
Arthropods 44, **45**, 66–72, **69, 71**
Aspirator 16, **16**
Aviary **9**

Bacteria 24, **24**, 86, 140
Barnacles 20, 70, 141
Bass 81, 140
Bat 121, **121, 122**
Beamer aspirator 16, **16**
Bears 126–127, **126, 127**
Beaver 132, **133**, 140
Beetle **10**, 14, 67, 68, **69**, 138, 139
Berlese extractor 19, **19**, 70
Berry **36, 37**, 106, 107
Birds, 77–78, **78**, 103–119, **103–105, 112–119**, 120
  banding 107–108, **107**
  feeding 106–107
  game **103**, 113, **114**, 116, **117**
  houses 106, 108, **108**, 109
  of prey 111, **105, 112**
  perching birds 118, **119**
  scavengers 112, **113**
  things to do with 105–110
  water birds 105, **105**, 109–110, 114–116
  watching 106–107, **107**
Bison 129, **129**, 138
Bivalve mollusks 64, **65**
Blackbird, Red-winged **119**, 138
Blind **106**
Bone 9, **11**, 105, 109–110, **111**, 154
Botanical Gardens 5, 8, **8**, 24
Bracken 34, **34**
Bryophyta 24
Bryozoa 44, **45**, 60, **60**

Bugs 66, 68, **69**
Butterfly 17, 66, 68, **69**
  nets **14**, 15

Calcite **145**
Caging 136, **136**
Cardinal 107, 118, **119**
Carp **80**, 81, 140
Caterpillar, preserving of 154
Centipede 44, **71**
Chickadee 106, **119**
Chipmunk **130**, 131
Chiton 63, 141
Chordates 44, **45**, 75–136
Club Moss 24, 33, **33**
Cnidaria 44, **49–53**, 140
Cod 80
Coelenterates 44, **45**, 49–53, **49–53**
Collection areas
  Desert 142, **142**
  Forest 139, **139**
  Fresh Water 140, **140**
  Grassland 138, **138**
  Seashore 141, **141**
Coot 114, 115, **115**, 138
Coral 44, 49, 52–53, **53**
Coyote **138**
Crab 70, **71**, 141, 145
Crocodilians 99–100, **100**
Crow 112, **113**
Crustacea 44, 70, **71**, 141

Daisy **39**, 138
Dandelion **25, 39**
Deer 127, **128**, 138, 139
Desert 142, **142**
Dinosaur fossils 144
Displays 150–151
Dogfish 78, **79**
Dog violet **39**
Dragonfly **11**, 17, **69**, 138
Ducks 116, **117**, 138

Eagle 111, **112**
Earthworm 44, 61, **62**
Echinoderm 44, **45**, 72–74, **72–74**, 146
Eel 81, **81**
Elk 127, **128**, 138
Extraction funnel 19, **19**

Fern 24, **24**, 34–35, **34, 35**, 139
  Cinnamon 34, **35**
  Sensitive 35, **35**
Fern allies 24, 33, **33**
  Field Horsetail **33**
  Running Ground Pine **33**
Fish 77, **77**, 78–86, **79–83**, **85**, 120, 138
Flatworm 44, **45**, 54–55, **54, 55**
Flounder 82, **82**
Flower 24, **25, 37**, 38–40, **38, 39**, 138, 139, 140
Fluke 44, 54, 55, **55**
Footwear 20, **22**

Forest 139, **139**
Fossil 10–11, **10, 22**, 23, 103, 143–146, **143, 144**
Fox 124, **124**, 138
Freeze Drying 43
Fresh-water 26, 48, 51–52, 54, 58, 61, 62, 70, 83–86
Frog **87**, 87–89, **90**, 92–93, 138, 140
Fungi 24, **24**, 28–29, **28–29**, 42, 139

Gabbro **149**
Geese **115**, 115–116
Granite **150**
Grass 25, 38, 138
Grassland 138, **138**
Grosbeak 118, **119**
Ground ivy **39**
Grouse 113, **114**
Gull 105, **105**, 110, 114, 141
Gymnosperm 36–37, **36, 37**

Hand lens 14, **21**, 22
Hanging Gardens of Babylon 4, **4**, 5
Hare 123, **123**
Hemichordates 44, **45**, 74
Herbarium 24–25
Heron 115, **115**
Heron pellets 105, 109–110
Herring 80
Horned Lizard **95**, 96, 142
Horsetail 24, 33, **33**
Hunter, John 7
Hunterian collection 7
Hydra 52–53, 140
Hydroid colony 49, **49**, 144

Igneous rocks 10
Illumination of aquaria 85
Insects 44, **45**, 66–69, **69**, 138, 139, 140, 142, 152–154
  collecting 15–17, 19, 66 68
  fossil **143**
  mounting 9, 67, 68, **153**
Invertebrates 20, 44, 145
Iris **38**
Iron pyrites **146**

Jay 107, 112, **113**, 139
Jellyfish 44, 49–52, **49–51**, 141
Juniper **36**

Kangaroo Rat **142**
Kestrel 111
Killdeer 114, **115**
Killing agents 16
Killing jars 16, **17**
Kit fox **142**

Labeling 21, 41, 149–150
Lady Slipper **38**, 139
Lamprey 78, **79**, 140
Lancelet 75, 76, **76**

157

Larch **36**
Leaf
    fossils **144**
    printing 154–155
Leech 44, 62, **62**
Lichen 20, 24, **25**, 28, 30–31, **31**, 139
Light trap 67
Limpet 63, **63**, 141
Liverwort 24, **24**, **30**, 32, **33**
Lizard 94–97, **95**, **96**, 102, 142

Mackerel shark 78, **79**
Mallard 116, **117**
Mammals 18–19, 78, **78**, 120–136, 154
    Carnivores 124–127, **124–127**
    Flying 121, **121**, **122**
    Hares and Rabbits 123, **123**
    Hoofed 127–129, **128**, **129**
    Insectivores 122, **122**
    Marsupials 120, **120**
    Rodents 130–132, **130**, **132**, **133**
    Trapping **18**, 18–19, 135
Mammoth 143
Metamorphic rocks 11
Millipede **71**
Minerals 10–11, 20, 23, 143, 145, 146–148
Minnow **80**, 81, 83, 140
Models 151–152
Mole 122, **122**
Mollusk 44, **45**, 62–65, **63**, **64**, **65**, 141
Moss 24, **24**, **30**, 31, 139
Moth 15, 17, 66, 68
    larvae **11**
Mold 13, 24, **25**, 28, 30
Moose 127, **128**
Mounting
    animals 110, 133, 152
    feathers 109
    insects 67, 68
    plants 25, 26, 40, 154, 155
    snake skin 102, **102**
Mouse, house **130**, 131
Museum 8, **9**, 27
Mussel 44, 63, 86, 141

National Audubon Society 104, 107
National Parks 8
National Wildlife Society 104
Nature Conservancy 104
Nestboxes 108, **108**, **109**
Nets 14–15, **14**, **15**
Newt 90–91, **91**, 93–94
Notebook 22, 43
Nummulite 46, **47**

Octopus 44, 62, 63, 65, **66**
Onyx **150**
Opossum **120**, 120, 139
Oriole 118, **119**
Owl **105**, 111, 139
    pellets 9, 109–110, **111**

Oyster 62, 63

Parasite 46, 150
Pebbles **147–148**, 147, 148–149
Photography 10, 43, 102, 133
Pink **25**
Planaria 44, 54
Plankton 15, 86, 137, **137**
Plants 8–9, 12, 24–43, **24–43**, 138–142, **144**, 154, 155
    classification 24–25
    drying 25, 27, 30, 40, **41**
    preserving 29, 30, 43, 36, 40–43, **152**, 154–155
    pressing 9, 12, **12**, **40**, 40–41, 154
Plaster casting 135, **135**, **150**, 151–152
Plastic bags 14, 21, 30
Polychaeta 44, 60–61, **61**
Pondweed 26, **26**
Pooter, see aspirator
Porbeagle **79**
Porcupine 131–132, **132**, 139
Porifera 44, 48, **48**
Portuguese Man-of-War 50, **50**
Prairie Dog **138**
Preservation of
    animals 48, 51–52, 61, 63, 64, 65, 67, 68, 70, 72, 75, 76, 102, 133–134, 137, 152, **152**
    fossils 146
    freeze drying 43
    green color in plants 42, **43**
    plants 25, 29, 30, 32, 40–43, **152**, 154–155
    rocks and minerals 147–148
Pressing Plants 12–13, **12**
Protozoa 44–47, **45**, **46**
Psota aspirator 16, **16**
Pteridophyta 24

Quail 113, **114**, 142
Quillwort 33

Rabbit 123, **123**, 133, 134, 139, 142
Ragged Robin **38**
Rat 130, **131**, 140
Ray 78
Reindeer **31**
Reptiles 77, **77**, 94–102
Ribbon worms 55, **56**
Riker mounts 67
Robin 104, 107
    nestbox **107**
Rocks 10–11, **10**, **11**, 20, 143–149, **143–149**, 152
Roman circus 4, **5**
Rotifer 44, **45**, 58–59, **59**
Round Worms 44, **45**, 56–58, **56**, **57**, **58**
Ruysch, Frederick **6**, 6–7

Safety precautions 23
Salamanders 90–95, **91–93**
Sand bath 40, **41**
Sand Dollars 73–74
Scavengers 112, **113**
Schist **149**
Scots Pine **36**
Sea anemone 44, 49, 51–53, **52**
Sea Cucumber 44, 72, 74, **74**
Sea Gooseberry, 44, 49, **52**, 53
Seahorse 82, **82**, 86
Seal 125, **125**
Sea Lily 44, 74
Seashore 10, 11, 20, 49, 60, 61, 63, 141, **141**, 152
Sea-squirt 44, **45**, 75, 76, **76**
Sea urchin 44, 72, 73, **73**, 86, **114**
Sea water, artificial 86
Seaweed **11**, 24, 26–27, **27**, 141
Sedimentary rocks 10
Seed plants 24, 36–39, 155
Serpentine **150**
Shell **10**, **22**
Shrew 18, 122, **122**
Shrimp 70, **71**, 86
Sierra Club 104
Skate 78, **79**
Skeleton 110, 133–134, **134**
Skins 102, **102**, 134

Tapeworm 44, **55**, 55
Thrush 104, 107
Titmouse **104**, 107
Toad 87–88, **89**, 92–94, 139, 140
Toadstool 24, **24**
Traps
    insect 19, **19**
    mammal **18**, 18–19, 135
Trilobite **144**
Trout 80, **83**, 84, 140
Trout Lily **38**
Turtles 100–102, **100**, **101**, 138

Vasculum 13, **13**
Vertebrates **8**, 44, 77–136, **77–83**, **85**, **87**, **89–93**, 95–101, **103–105**, **112–130**, **132–134**, **138–142**
Vinegar Eel 56, **57**
Violet **25**
Vivarium 94, **94**, 102

Warbler 118, **119**, 138
Weasel 124, **124**
Woodchuck 131, **132**
Woodland 139, **139**
Woodpecker **116**, 118
Wren **104**
    nest box **109**, 118

Yew 36, **37**

Zincite **146**
Zoological Gardens 8

OTHER TITLES IN THE SERIES

The GROSSET ALL-COLOR GUIDES provide a library of authoritative information for readers of all ages. Each comprehensive text with its specially designed illustrations yields a unique insight into a particular area of man's interests and culture.

BIRD BEHAVIOR	FLAGS OF THE WORLD
PREHISTORIC ANIMALS	GUNS
BIRDS OF PREY	ARMS AND ARMOR
WILD CATS	ARCHITECTURE
MAMMALS OF THE WORLD	MATHEMATICS
SNAKES OF THE WORLD	ELECTRONICS
FISHES OF THE WORLD	COMPUTERS AT WORK
MONKEYS AND APES	ATOMIC ENERGY
HORSES AND PONIES	MICROSCOPES AND
FLOWER ARRANGING	MICROSCOPIC LIFE
DOGS, SELECTION-CARE-	WEATHER AND
TRAINING	WEATHER FORECASTING
CATS, HISTORY-CARE-	PORCELAIN
BREEDS	VICTORIAN FURNITURE AND
TROPICAL FRESHWATER	FURNISHINGS
AQUARIA	SAILING SHIPS AND
NATURAL HISTORY	SAILING CRAFT
COLLECTING	TRAINS
FOSSIL MAN	VETERAN AND
ARCHEOLOGY	VINTAGE CARS
THE HUMAN BODY	WARSHIPS
EVOLUTION OF LIFE	AIRCRAFT
THE PLANT KINGDOM	ROCKETS AND MISSILES
THE ANIMAL KINGDOM	ASTRONOMY
HOUSE PLANTS	EXPLORING THE PLANETS
SEA BIRDS	MYTHS AND LEGENDS OF
BUTTERFLIES	ANCIENT GREECE
SEA SHELLS	MYTHS AND LEGENDS OF
TROPICAL MARINE	ANCIENT EGYPT
AQUARIA	DISCOVERY OF AFRICA
MILITARY UNIFORMS, 1686-1918	DISCOVERY OF NORTH AMERICA